T0384047

# Applied Safety
# for Engineers

# Applied Safety
# for Engineers

## Systems and Products

B.S. Dhillon

## CRC Press
Taylor & Francis Group
Boca Raton  London  New York

CRC Press is an imprint of the
Taylor & Francis Group, an **informa** business

First edition published 2022
by CRC Press
6000 Broken Sound Parkway NW, Suite 300, Boca Raton, FL 33487-2742

and by CRC Press
2 Park Square, Milton Park, Abingdon, Oxon, OX14 4RN

© 2022 B. S. Dhillon

CRC Press is an imprint of Taylor & Francis Group, LLC

**Library of Congress Cataloging-in-Publication Data**

Names: Dhillon, B. S. (Balbir S.), 1947- author.
Title: Applied safety for engineers : systems and products / B.S. Dhillon.
Description: First edition. | Boca Raton, FL : CRC Press, 2022. | Includes
bibliographical references and index.
Identifiers: LCCN 2021035677 (print) | LCCN 2021035678 (ebook) | ISBN
9781032080987 (hbk) | ISBN 9781032080994 (pbk) | ISBN 9781003212928
(ebk)
Subjects: LCSH: Industrial safety.
Classification: LCC T55 .D4848 2022 (print) | LCC T55 (ebook) | DDC
620.8/6--dc23
LC record available at https://lccn.loc.gov/2021035677
LC ebook record available at https://lccn.loc.gov/2021035678

ISBN: 978-1-032-08098-7 (hbk)
ISBN: 978-1-032-08099-4 (pbk)
ISBN: 978-1-003-21292-8 (ebk)

DOI: 10.1201/9781003212928

Typeset in Times
by Deanta Global Publishing Services, Chennai, India

*This book is affectionately dedicated to the memory of my late father-in-law*

*Jagir S. Grewal, late mother-in-law Raj K. Grewal, and father-in-law's late brother Dr. Ranjit S. Grewal.*

# Contents

..

# Preface

Nowadays, engineering systems/products are an important element of the world economy, and each year a vast sum of money is spent to develop, manufacture, operate, and maintain various types of engineering systems around the globe. The safety of these systems/products has become important than ever before because of their increasing complexity, sophistication, and non-specialist users. Global competition and other factors are forcing manufacturers to produce highly safe engineering systems/products.

It means that there is a definite need for safety and other professionals to work closely during design and other phases. To achieve this goal, it is essential that they have an understanding of each other's speciality/discipline to a certain degree. At present to the best of the author's knowledge, there is no book that is specifically concerned with applied safety in a wide range of areas. It means, at present, to gain knowledge of each other's specialities, these professionals must study various books, reports, and articles on each of the topics in question. This approach is time consuming and rather difficult because of the specialized nature of the material involved.

Thus, the main objective of this book is to combine a wide range of applied safety topics into a single volume and to treat the material covered in such a manner that the reader requires no previous knowledge to understand it. The sources of most of the material presented are given in the reference section at the end of each chapter. This will be useful to readers if they desire to delve more deeply into a specific area or topic. At appropriate places, the book contains examples along with their solutions, and at the end of each chapter there are numerous problems to test the reader's comprehension in the area.

The book is composed of 15 chapters. Chapter 1 presents various introductory aspects of applied safety including safety application areas, useful sources for obtaining information on safety, and safety standards. Chapter 2 reviews mathematical concepts considered useful to understand subsequent chapters. Some of the topics covered in the chapter are arithmetic mean and mean deviation, Boolean algebra laws, probability properties, probability distributions, and useful mathematical definitions. Chapter 3 presents various introductory aspects of safety.

Chapter 4 presents a number of methods considered useful to analyze engineering systems safety. These methods are hazards and operability analysis, failure modes and effect analysis, fault tree analysis, Markov method, control charts, interface safety analysis, preliminary hazard analysis, root cause analysis, technic of operation review, and job safety analysis. Chapter 5 is devoted to robot safety. Some of the topics covered in the chapter are robot safety-related problems and hazards, robot manufacturers' and users' roles in safety, common robot safety-related features and their functions, and robot safeguard approaches. Chapter 6 presents various important aspects of safety in nuclear power plants. Some of the topics covered in the chapter are safety objectives of nuclear power plants, nuclear power plant fundamental safety principles, nuclear power plant-specific safety principles, safety management in nuclear power plant design, and deterministic safety analysis for nuclear power plants.

Chapter 7 is devoted to medical systems safety. Some of the topics covered in the chapter are medical systems safety-related facts and figures, safety in medical device life cycle, software-related issues in cardiac rhythm management product safety, and medical system safety analysis methods. Chapter 8 presents various important aspects of airline and ship safety. Some of the topics covered in the chapter are United States airline-associated fatalities and accident rates, aircraft accidents during flight phases and causes of airplane crashes, worldwide airline accident analysis, noteworthy marine accidents, global maritime distress safety system, and ship safety assessment. Chapter 9 is devoted to rail safety. Some of the topics covered in the chapter are causes of railway-associated accidents and incidents, general categories of rail accidents by causes and effects, telescoping-associated railway accidents, railroad tank car safety, and methods for performing rail safety analysis.

Chapter 10 presents various important aspects of truck and bus safety. Some of the topics covered in the chapter are top truck and bus safety-related issues, most-cited truck safety-associated problems, safety culture in the trucking industrial sector, safety-associated truck inspection tips, bus and coach occupant fatalities and serious injuries, transit bus safety and key design-associated safety feature areas, and vehicle safety data sources. Chapter 11 is devoted to mining equipment safety. Some of the topics covered in the chapter are mining equipment safety-related facts and figures, main causes of mining equipment accidents, programmable-electronic-associated mishaps and lessons learned, equipment fire-associated mining accidents and mining equipment fire ignition sources, useful guidelines for improving electrical safety in mines, human factors-associated design tips for safer mining equipment, and methods for performing mining equipment safety analysis. Chapter 12 presents various important aspects of programmable electronic mining system safety. Some of the topics covered in the chapter are programmable electronics usage trends in the mining industry, lessons learned in addressing programmable electronic mining systems safety, methods for performing programmable electronic mining systems hazard and risk analysis, and sources for obtaining programmable electronic mining system safety-related information.

Chapter 13 is devoted to safety in the offshore oil and gas industry. Some of the topics covered in the chapter are offshore industrial sector-related risk picture, offshore worker situation awareness concept, studies and their results, offshore industry deadliest accidents' case studies, and offshore industry accident reporting approach and offshore accident-associated causes. Chapter 14 presents various important aspects of software safety. Some of the topics covered in the chapter are software safety-related facts, figures, and examples; software safety classifications; basic software system safety-associated tasks; useful software safety design-related guidelines; software hazard analysis methods; and software standards. Finally, Chapter 15 is devoted to safety in engineering maintenance. Some of the topics covered in the chapter are facts, figures, and examples; maintenance safety-related problems' causes; good safety-associated practices during maintenance work; guidelines for equipment designers to improve safety in maintenance; maintenance safety-associated questions for engineering equipment manufacturers; and mathematical models.

This book will be useful to many individuals including design engineers, system engineers, safety professionals, researchers and instructors of safety, engineering administrators, graduate and senior undergraduate students in the area of engineering, and engineers at large.

The author is deeply indebted to many individuals including family members, colleagues, friends, and students for their invisible inputs. The invisible contributions of my children are also appreciated. Last, but not least, I thank my wife, Rosy, my other half and friend, for typing this entire book and for timely help in proofreading.

**B. Dhillon**
*University of Ottawa*

# Author

**Dr. B.S. Dhillon** is a professor of Engineering Management in the Department of Mechanical Engineering at the University of Ottawa. He has served as a chairman/director of the Mechanical Engineering Department/Engineering Management Program for over ten years at the same institution. He is the founder of the probability distribution named *Dhillon Distribution/Law/Model* by statistical researchers in their publications around the world. He has published over 377 (i.e., 224 (70 single-authored + 150 co-authored) journals and 153 conference proceedings) articles on reliability engineering, maintainability, safety, engineering management, etc. He is or has been on the editorial boards of 14 international scientific journals. In addition, Dr. Dhillon has written 49 books on various aspects of health care, engineering management, design, reliability, safety, and quality published by Wiley (1981), Van Nostrand (1982), Butterworth (1983), Marcel Dekker (1984), Pergamon (1986), etc. His books are being used in over 100 countries, and many of them are translated into languages such as German, Russian, Chinese, and Persian (Iranian).

He has served as general chairman of two international conferences on reliability and quality control held in Los Angeles and Paris in 1987. Professor Dhillon has also served as a consultant to various organizations and bodies and has many years of experience in the industrial sector. At the University of Ottawa, he has been teaching reliability, quality, engineering management, design, and related areas and he has also lectured in over 50 countries, including delivering keynote addresses at various international scientific conferences held in North America, Europe, Asia, and Africa. In March 2004, Dr. Dhillon was a distinguished speaker at the Conference/Workshop on Surgical Errors (sponsored by White House Health and Safety Committee and Pentagon), held at the Capitol Hill (One Constitution Avenue, Washington, D.C.).

Professor Dhillon attended the University of Wales, where he received a BS in electrical and electronic engineering and an MS in mechanical engineering. He received his PhD in industrial engineering from the University of Windsor.

# Author

Dr. A.S. Dhillon is a ... professor of Engineering ... in the Department of Mechanical Engineering ...

# 1 Introduction

## 1.1 SAFETY HISTORY

The history of safety may be traced back to 2000 BC when the ancient Babylonian ruler Hammurabi developed a code known as the Code of Hammurabi. This code contained clauses, directly or indirectly, concerning areas such as injuries, allowable fees for physicians, and monetary damages assessed against those who caused injury to others [1,2]. In the 4th century BC, Hippocrates, a Greek doctor, highlighted lead poisoning, and a Roman called Pliny the Elder (AD 23–79) pointed to fumes from lead as well as the dust from mercury ore grinding and recommended the wearing of protective masks by involved workers [3,4].

In modern times, in 1868, a patent was awarded for the first barrier safeguard and in 1877, the Massachusetts legislature passed a law requiring appropriate safeguards on hazardous machinery [1]. In 1893, the United States Congress passed the Railway Safety Act and in 1913, the National Council of Industrial Safety was formed and changed its name to the National Safety Council in 1915 [1,5]. In 1969, the United States Department of Defense released MIL-STD-882 titled "System Safety Program for Systems and Associated Subsystems and Equipment: Requirements for" and in 1970, the United States Congress passed the Occupational Safety and Health Act.

Additional information on the safety history is available in Ref. [6].

## 1.2 NEED FOR SAFETY AND ENGINEERING SAFETY GOALS

Safety has become a very important issue because each year a large number of people die and get seriously injured due to workplace-related and other accidents. For example, in the United States alone for the year 1996, as per the National Safety Council, there were 93,400 fatalities and a large number of disabling injuries due to accidents with a total loss of US$121 billion [7]. In addition, factors such as lawsuits, governmental regulations, and public pressures also play a key role in demanding the need for better safety.

There are many engineering safety goals and some of these are as follows:

- Eliminate or control hazards.
- Reduce accidents.
- Maximize public confidence in regard to safety.
- Develop new methods and techniques for improving safety.
- Maximize returns on safety-related efforts.

DOI: 10.1201/9781003212928-1

## 1.3   SAFETY APPLICATION AREAS

Ever since the inception of the safety field, the safety discipline has branched into many application areas such as follows:

- **Robot safety:** This is an emerging new area of the application of basic safety principles to robot-related problems over the years, many publications on the subject have appeared, including two textbooks [8,9].
- **Medical systems safety:** This is also an emerging new area of the application of basic safety principles to medical system-associated problems. Over the years, many publications on the topic have appeared, including two textbooks which present some material on the topic [10,11].
- **Airline and ship safety:** This is also an emerging new area of the application of basic safety principles to airline- and ship-related problems. Over the years, many publications on the topic have appeared, including one textbook which presents some material on the topic [12].
- **Rail safety:** This is also an emerging new area of the application of basic safety principles to rail safety-related problems. Over the years many publications on the topic have appeared, including one textbook which presents some material on the topic [12].
- **Mining equipment safety:** This is also an emerging new area of the application of basic safety principles to mining equipment-related problems. Over the years, many publications on the topic have appeared, including two textbooks which present some material on the topic [13,14].
- **Safety in offshore oil and gas industry:** This is also an emerging new area of the application of basic safety principles in offshore oil and gas industry-related problems. Over the years, many publications on the topic have appeared, including one textbook which presents some material on the topic [15].
- **Software safety:** This is a very important emerging area of safety as the use of computers and other electronic devices is increasing at an alarming rate. Over the years, many publications on the topic have appeared, including two textbooks which present some material on the topic [6,16].
- **Truck and Bus safety:** This is also an emerging new area of the application of basic safety principles to truck and bus safety-associated problems. Over the years many publications on the topic have appeared, including one textbook which presents some material on the topic [12].
- **Safety in nuclear power plants:** This is also an emerging area of the application of basic safety principles in nuclear power plant-related problems. Over the years many publications on the topic have appeared, including one textbook which presents some material on the topic [17].

## 1.4   SAFETY FACTS AND FIGURES

Some of the facts and figures directly or indirectly concerned with the subject of safety are as follows:

- In 1995, work-associated accidents cost the United States economy around US$75 billion [18].
- In 2000, there were a total of 97,300 unintentional injury-related deaths in the United States and their total cost was estimated to be approximately US$512.4 billion [19].
- In the European Union, approximately 5,500 persons are killed due to workplace-related accidents each year [20].
- During the period 1978–1987, there were ten robot-associated fatal accidents in Japan [21].
- In 1969, the U.S. Department of Health, Education, Welfare special committee reported that over a period of ten years in the United States, there were approximately 10,000 medical device-related injuries and 731 resulted in fatalities [22,23].
- In a typical year, around 35 million work hours are lost due to accidents in the entire United States [24].
- Two patients died and a third was severely injured due to a software error in a computer-controlled therapeutic radiation machine known as Therac 25 [25–27].
- As per Ref. [28], the annual cost of world road crashes is over US$500 billion.
- In 1986, the Space Shuttle Challenger exploded and all its crew members were killed [6,29].
- In 1986, a nuclear reactor in Chernobyl, Ukraine, exploded and directly or indirectly caused approximately 10,000 fatalities [6,29].

## 1.5  TERMS AND DEFINITIONS

There are a large number of terms and definitions used in the area of safety. Some of these are as follows [1,18,30–35]:

- **Safety:** This is the conservation of human life and the prevention of damage to items as per mission requirements.
- **Unsafe condition:** This is any condition, under the right set of conditions, that will lead to an accident.
- **Hazard:** This is the source of energy and the physiological and behavioural factors which, when uncontrolled, result in harmful occurrences.
- **Accident:** This is an event that involves damage to a certain system that suddenly disrupts the current or potential system output.
- **Safeguard:** This is a barrier guard, device, or procedure developed for protecting people.
- **Safety management:** This is the accomplishment of safety through the efforts of others.
- **Unsafe act:** This is an act that is not safe for an employee/individual.
- **Safety assessment:** This is a qualitative/quantitative determination of safety.

- **Safety process:** This is a series of procedures followed to enable all safety-related requirements of an item/system to be identified and satisfied.
- **Safety plan:** This is the implementation details of how the safety-related requirements of the project will be achieved.
- **Unsafe behaviour:** This is the manner in which an individual carries out actions that are considered unsafe to himself/herself or other people.
- **Safe:** This is protected against any possible hazard.
- **Software safety:** This is the freedom from software-associated hazard.
- **Software hazard:** This is a software condition prerequisite to an accident.
- **Hazard control:** This is a means of reducing the risk from exposure to a perceived hazard.
- **Injury:** This is a wound or other specific/certain damage.
- **Accident report:** This is a document that records the findings of an accident investigation, the accident cause/causes, and the recommended actions.

## 1.6    USEFUL SOURCES FOR OBTAINING INFORMATION ON SAFETY

This section lists books, journals, standards, organizations, and sources considered useful for obtaining information directly or indirectly concerned with applied safety.

### 1.6.1    BOOKS

- Handley, W., *Industrial Safety Handbook*, McGraw Hill Book Company, London, 1969.
- Stephans, R.A., Taslo, W.W., Eds., *System Safety Analysis Handbook*, System Safety Society, Irvine, CA, 1993.
- Hall, S., *Railway Accidents*, Ian Allan Publishing, Shepperton, U.K., 1997.
- Spellman, F.R., Whiting, N.E., *Safety Engineering: Principles and Practice, Government Institutes*, Rockville, Maryland, 1999.
- Manuele, F.A., *On the Practice of Safety*, Van Nostrand Reinhold Co., New York, 1993.
- Tarrants, W.E., *The Measurement of Safety Performance*, Garlnd STPM Press, New York, 1980.
- Dhillon, B.S., *Engineering Safety: Fundamentals, Techniques, and Applications*, World Scientific Publishing, River Edge, New Jersey, 2003.
- Heinrich, W.H., *Industrial Accident Prevention*, 3rd Ed., McGraw Hill Book Company, New York, 1950.
- Asfahl, C.R., *Industrial Safety and Health Management*, Prentice Hall, Inc., Englewood Cliffs, New Jersey, 1990.
- Gloss, D.S., Wardle, M.G., *Introduction to Safety Engineering*, John Wiley and Sons, New York, 1984.
- Hammer, W., Price, D., *Occupational Safety Management and Engineering*, Prentice Hall, Upper Saddle River, New Jersey, 2001.

- Raouf, A., Dhillon, B.S., *Safety Assessment: A Quantitative Approach*, Lewis Publishers, Boca Raton, Florida, 1994.
- Leveson, N.G., *Safeware: System Safety and Computers*, Addison-Wesley, Reading, Massachusetts, 1995.
- Eastman, C., *Work Accidents and the Law*, New York Charities Publication Committee, New York, 1910.

### 1.6.2 JOURNALS

- *Nuclear Safety*
- *Journal of Safety Research*
- *Accident Analysis and Prevention*
- *Safety Management Journal*
- *Hazard Prevention*
- *Safety Science*
- *Journal of Occupational Accidents*
- *Safety and Health*
- *Professional Safety*
- *National Safety News*
- *Product Safety News*
- *Asia Pacific Air Safety*
- *International Journal of Reliability, Quality, and Safety Engineering*
- *Reliability Engineering and System Safety*
- *Risk Analysis*
- *Australian Safety News*
- *Accident Prevention*
- *Air Force Safety Journal*
- *Journal of Fire Safety*

### 1.6.3 STANDARDS

- **MIL-STD-58077**, Safety Engineering of Aircraft System, Associated Subsystem and Equipment-General Requirements, Department of Defense, Washington, D.C., 1967.
- **MIL-STD-882**, Systems Safety Program for System and Associated Subsystem and Equipment-Requirements, Department of Defense, Washington, D.C., 1969.
- **IEC 60950**, Safety of Information Technology Equipment, International Electrotechnical Commission, Geneva, Switzerland, 1999.
- **IEC 60880**, Software for Computers in the Safety Systems of Nuclear Power Stations, Internal Electrotechnical Commission, Geneva, Switzerland, 1986.
- **MIL-STD-23069**, Safety Requirements Minimum for Air Launched Guided Missiles, Department of Defense, Washington, D.C., 1965.
- **DEF STD 00-55-1**, Requirements for Safety-Related Software in Defense Equipment, Department of Defense, Washington, D.C., 1997.

- **IEC 601-1:1988**, Medical Electrical Equipment Part I: General Requirements for Safety, International Electrotechnical Commission, Geneva, Switzerland, 1988.
- **IEC 61508 SET**, Functional Safety of Electrical/Electronic/Programmable Electronic Safety-Related Systems-Parts 1-7, International Electrotechnical Commission, Geneva, Switzerland, 2000.
- **AECL CE-1001-STD (revision 2)**, Standard for Software Engineering of Safety Critical Software, Atomic Energy of Canada Limited, Ottawa, 1999.

### 1.6.4 ORGANIZATIONS

- The International Institute of Risk and Safety Management, 70 Chancellors Road, London, UK.
- World Safety Organization, P.O. Box No. 1, Lalong Laan Building, Pasay City, Metro Manila, The Philippines.
- British Safety Council, 62 Chancellors Road, London, UK.
- The American Society of Safety Engineers, 1800 E. Oakton Street, Des Plaines, Illinois, USA.
- System Safety Society, 14252 Culver Drive, Suite A-261, Irvine, California, USA.
- Institution of Occupational Safety and Health, The Grange Highfield Drive, Wigston, Leicestershire, UK.
- Occupational Safety and Health Administration (OSHA), U.S. Department of Labor, 200 Constitution Avenue, Washington, D.C., USA.
- Civil Aviation Safety Authority, North Bourne Avenue and Barry Drive Intersection, Canberra, Australia.
- Federal Railroad Administration, 4601 N. Fairfax Drive, Suite 1100, Arlington, Virginia, USA.
- Transportation Safety Board of Canada, 330 Spark Street, Ottawa, Ontario, Canada.
- Board of Certified Safety Professionals, 208 Burwash Avenue, Savoy, Illinois, USA.
- U.S. Consumer Product Safety Commission, Washington, D.C., USA.
- Pacific Safety Council, 7220 Trade, Suite 100, San Diego, California, USA.
- National Institute for Occupational Safety and Health (NIOSH), 200 Independence Avenue, SW Washington, D.C., USA.
- International Civil Aviation Organization, 999 University Street, Montreal, Quebec, Canada.

### 1.6.5 DATA INFORMATION SOURCES

- GIDEP Data, Government Industry Data Exchange Program (GIDEP) Operational Center, Fleet Missile Systems, Analysis, and Evaluation, Department of the Navy, Corona, California, USA.

- National Technical Information Service (NTIS), 5285 Port Royal Road, Springfield, Virginia, USA.
- National Safety Information Center, Oak Ridge National Laboratory, P.O. Box Y, Oak Ridge, Tennessee, USA.
- Health and Safety Executive Line, European Space Agency (ESP) Information Retrieval Service, Via Galileo 00044, Frascati, Rome, Italy.
- National Electronic Injury Surveillance System, US Consumer Product Safety Commission, 5401 Westbard Street, Washington, D.C., USA.
- Safety Research Information Center, National Safety Council, 444 North Michigan Avenue, Chicago, Illinois, USA.
- Computer Accident/Incident Report System, System Safety Development Center, EG8G, P.O. Box 1625, Idaho Falls, Idaho, USA.
- International Occupational Safety and Health Information, Center Bureau, International du Travail, CH-1211, Geneva, Switzerland.

## 1.7 SCOPE OF THE BOOK

Nowadays, engineering systems are a very important element of the world economy, and each year, a vast sum of money is spent on developing, manufacturing, operating, and maintaining various types of engineering systems around the globe. Global competition and other factors are directly or indirectly forcing manufacturers to produce highly safe engineering systems/products. Over the years, a large number of journal and conference proceedings articles, technical reports, etc., on the safety of engineering systems have appeared in the literature. However, to the best of the author's knowledge, there is no book that covers the topic of applied safety only within its framework. This is a significant impediment to information seekers on this topic because they have to consult various sources.

Thus, the main objectives of this book are as follows:

- To eliminate the need for professionals and others concerned with applied safety to consult various different and diverse sources in obtaining the desired information.
- To provide up-to-date information on the subject.

This book will be useful to many individuals including design engineers, system engineers, safety professionals concerned with engineering systems, engineering system administrators, researchers and instructors in the area of engineering systems, engineering graduate and senior undergraduate students, and engineers at large.

## 1.8 PROBLEMS

1. Define the following terms:
   - Safety
   - Safety process
   - Hazard

2. Write an essay on the history of safety.
3. What are the main goals of engineering safety?
4. Discuss the following safety application areas:
   - Robot safety
   - Medical systems safety
   - Software safety
5. List at least five important facts and figures concerned with safety.
6. List at least seven safety-related standards.
7. Define the following terms:
   - Accident
   - Unsafe behaviour
   - Safety management
8. Write down six sources for obtaining safety-related information.
9. List at least eight organizations that are specifically concerned with safety.
10. List at least ten journals considered most useful for obtaining engineering systems safety-related information.

## REFERENCES

1. Goestsch, D.L., *Occupational Safety and Health*, Prentice Hall, Englewood Cliffs, New Jersey, 1996.
2. Ladon, J., Ed., *Introduction to Occupational Health and Safety*, National Safety Council, Chicago, 1986.
3. *Pliny the Elder (AD 23–79) Historia Naturalis*, trans. by H. Rackham, Harvard University Press, Cambridge, Massachusetts, 1947.
4. Dhillon, B.S., *Engineering Design: A Modern Approach*, Richard D. Irwin, Chicago, 1996.
5. Hammer, W., Price, D., *Occupational Safety Management and Engineering*, Prentice Hall, Upper Saddle River, New Jersey, 2001.
6. Dhillon, B.S., *Engineering Safety: Fundamentals, Techniques, and Applications*, World Scientific Publishing, River Edge, New Jersey, 2003.
7. *Accidental Facts*, Report, National Safety Council, Chicago, Illinois, 1996.
8. Dhillon, B.S., *Robot Reliability and Safety*, Springer-Verlag, New York, 1991.
9. Dhillon, B.S., *Robot System Reliability and Safety*, CRC Press, Boca Raton, Florida, 2015.
10. Dhillon, B.S., *Medical Device Reliability and Associated Areas*, CRC Press, Boca Raton, Florida, 2000.
11. Dhillon, B.S., *Patient Safety: An Engineering Approach*, CRC Press, Boca Raton, Florida, 2011.
12. Dhillon, B.S., *Transportation Systems Reliability and Safety*, CRC Press, Boca Raton, Florida, 2011.
13. Dhillon, B.S., *Mining Equipment Reliability, Maintainability, and Safety*, Springer-Verlag, London, 2008.
14. Dhillon, B.S., *Mine Safety: A Modern Approach*, Springer-Verlag, London, 2010.
15. Dhillon, B.S., *Safety and Reliability in the Oil and Gas Industry: A Practical Approach*, CRC Press, Boca Raton, Florida, 2016.
16. Dhillon, B.S., *Computer System Reliability: Safety and Usability*, CRC Press, Boca Raton, Florida, 2013.

17. Dhillon, B.S., *Safety, Reliability, Human Factors, and Human Error in Nuclear Power Plants*, CRC Press, Boca Raton, Florida, 2018.

18. Spellman, F.R., Whiting, N.E., *Safety Engineering: Principles and Practice*, Government Institutes, Rockville, MD, 1999.

19. *Report on Injuries in America in 2000*, National Safety Council, Chicago, Illinois, 2000.

20. *How to Reduce Workplace Accidents*, European Agency for Safety and Health at Work, Brussels, Belgium, 2001.

21. Nagamachi, M., Ten Fatal Accidents Due to Robots in Japan, in *Ergonomics of Hybrid Automated Systems*, edited by W. Karwowski, et al., Elsevier, Amsterdam, 1988, pp. 391–396.

22. Banta, H.D., The Regulation of Medical Devices, *Preventive Medicine*, Vol. 19, 1990, pp. 693–699.

23. *Medical Devices, Hearings before the Sub-committee on Public Health and Environment, U.S. Congress Interstate and Foreign Commerce*, Serial No. 93-61, U.S. Government Printing Office, Washington, D.C., 1973.

24. *Accidents Facts*, National Safety Council, Chicago, 1990–1993.

25. Gowen, L.D., Yap, M.Y., Traditional Software Development's Effects on Safety, Proceedings of the 6th Annual IEEE Symposium on Computer-Based Medical Systems, 1993, pp. 58–63.

26. Joyce, E., Software Bugs: A Matter of Life and Liability, *Datamation*, Vol. 33, No. 10, 1987, pp. 88–92.

27. Schneider, P., Hines, M.L.A., Classification of Medical Software, Proceedings of the IEEE Symposium on Applied Computing, 1990, pp. 20–27.

28. Odero, W., Road Traffic Injury Research in Africa: Context and Priorities, Paper presented at the Global Forum for Health Research Conference (Forum 8), November 2004. Available from the School of Public Health, Moi University, Eldoret, Kenya.

29. Schlager, N., *Breakdown: Deadly Technological Disasters*, Visible Ink Press, Detroit, 1995.

30. Roland, H.E., Moriarty, B., *System Safety Engineering and Management*, John Wiley and Sons, New York, 1983.

31. Grimaldi, J.V., Simonds, R.H., *Safety Management*, Richard D. Irwin, Chicago, 1989.

32. *Dictionary of Terms Used in the Safety Profession*, 3rd Ed., American Society of Safety Engineers, Des Plaines, Illinois, 1988.

33. MIL-STD-498, *Software Development and Documentation*, Department of Defense, Washington, D.C., 1994.

34. Meulen, M.V.D., *Definitions for Hardware and Software Safety Engineers*, Springer-Verlag, London, 2000.

35. IEEE-STD-1228, *Standard for Software Safety Plans*, Institute of Electrical and Electronic Engineers, New York, 1994.

# 2 Basic Mathematical Concepts

## 2.1 INTRODUCTION

Just like in the development of other areas of engineering, mathematics has also played an important role in the development of the safety field. The history of mathematics may be traced back to the development of our current number symbols often referred to as "Hindu-Arabic numeral system" in the published literature [1]. Among the first evidences of the use of these numerals/symbols is found on stone columns erected by the Scythian Emperor of India named Asoka, around 250 BC[1].

The earliest reference to the probability concept may be traced back to the gambling manual written by Girolamo Cardano (1501–1576) [1,2]. However, Pierre Fermat (1601–1665) and Blaise Pascal (1623–1662) were the first two individuals who solved independently and correctly the problem of dividing the winnings in a game of chance. Boolean algebra, which plays a pivotal role in modern probability theory, is named after the mathematician George Boole (1815–1864), who published, in 1847, a pamphlet titled *The Mathematical Analysis of Logic: Being an Essay towards a Calculus of Deductive Reasoning* [1,3]. Additional information on the history of mathematics and probability is available in Refs. [1,2].

This chapter presents basic mathematical concepts that will be useful to understand subsequent chapters of this book.

## 2.2 ARITHMETIC MEAN AND MEAN DEVIATION

A given set of safety data is useful only if it is analyzed properly. More specifically, there are certain characteristics of the data that are quite helpful in describing the nature of a given data set, thus making better-related decisions. Thus, this section presents two statistical measures considered useful to study engineering systems safety-related data [4,5].

### 2.2.1 ARITHMETIC MEAN

Generally, this is simply referred to as the mean and is defined by

$$m = \frac{\sum_{j=1}^{k} y_i}{k} \tag{2.1}$$

DOI: 10.1201/9781003212928-2

where
  $m$ is the mean value (i.e., arithmetic mean).
  $y_j$ is the data value $j$, for $j = 1,2,3,\dots,k$.
  $k$ is the total number of data values.

### Example 2.1

Assume that the inspection department of an engineering systems manufacturing company inspected six identical systems and discovered 2, 4, 3, 6, 1, and 8 safety-related problems. Calculate the mean or average number of safety problems per system (i.e., arithmetic mean).
  By substituting the specified data values into Equation (2.1), we obtain:

$$m = \frac{2+4+3+6+1+8}{6} = 4$$

Thus, the mean or average number of safety-related problems per system (i.e., arithmetic mean) is 4.

### 2.2.2 Mean Deviation

This is a measure of dispersion, and its value indicates the degree to which a given set of data tend to spread about a mean value. Mean deviation is expressed by:

$$MD = \frac{\sum_{j=1}^{k}|y_j - m|}{k} \tag{2.2}$$

where
  $MD$ is the mean deviation.
  $k$ is the total number of data values.
  $m$ is the mean value (i.e., arithmetic mean).
  $y_j$ is the data value $j$, for $j = 1,2,3,\dots,k$.
  $|y_j - m|$ is the absolute value of the deviation of $y_i$ from $m$.

### Example 2.2

Calculate the mean deviation of the dataset given in Example 2.1.
  In Example 2.1, the calculated mean value of the given data set is four safety-related problems per system. Thus, by inserting this calculated value and the given data values into Equation (2.2), we obtain:

$$MD = \frac{\left[|2-4|+|4-4|+|3-4|+|6-4|+|1-4|+|8-4|\right]}{6}$$

$$= \frac{2+0+1+2+3+4}{6}$$

$$= 2$$

Thus, the mean deviation of the dataset given in Example 2.1 is 2.

## 2.3  BOOLEAN ALGEBRA LAWS

Boolean algebra is used to a degree in engineering systems-associated studies and is named after George Boole (1815–1864), its founder. Some of the Boolean algebra laws are as follows [3,6]:

- Commutative law

$$X + Y = Y + X \tag{2.3}$$

$$X.Y = Y.X \tag{2.4}$$

where
X is an arbitrary set or event.
Y is an arbitrary set or event.
Dot (.) denotes the intersection of sets. (Note that sometimes Equation (2.4) is written without the dot, but it still conveys the same meaning.)
+ is the union of sets or events.
- Idempotent law

$$X + X = X \tag{2.5}$$

$$XX = X \tag{2.6}$$

- Absorption law

$$X + XY = X \tag{2.7}$$

$$Y(X + Y) = Y \tag{2.8}$$

- Associative law

$$(XY)Z = X(YZ) \tag{2.9}$$

$$(X + Y) + Z = X + (Y + Z) \tag{2.10}$$

where Z is an arbitrary set or event.
- Distributive law

$$X(Y + Z) = XY + XZ \tag{2.11}$$

$$(X + Y)(X + Z) = X + YZ \tag{2.12}$$

## 2.4  PROBABILITY DEFINITION AND PROPERTIES

Probability is defined as follows [7]:

$$P(X) = \lim_{m \to \infty} \left( \frac{M}{m} \right) \tag{2.13}$$

where
  $P(X)$ is the probability of occurrence of event $X$.
  $M$ is the number of times event $X$ occurs in the $m$-repeated experiments.

Some of the basic properties of probability are as follows [7,8]:

- The probability of occurrence of event X is

$$0 \le P(X) \le 1 \tag{2.14}$$

- The probability of occurrence and non-occurrence of event X is always

$$P(X) + P(\bar{X}) = 1 \tag{2.15}$$

where
  $P(X)$ is the probability of occurrence of event $X$.

  $P(\bar{X})$ is the probability of non-occurrence of event $X$.

- The probability of the union of m mutually exclusive events is

$$P\left(X_1 + X_2 + - - - + X_m\right) = \sum_{i=1}^{m} P(X_i) \tag{2.16}$$

where
  $P(X_i)$ is the probability of occurrence of event $X_i$, for $i=1,2,3,\ldots,m$.

- The probability of the union of $m$ independent events is

$$P(X_1 + X_2 + - - - - + X_m) = 1 - \prod_{i=1}^{m}(1 - P(X_i)) \tag{2.17}$$

- The probability of an intersection of $m$ independent events is

$$P(X_1 X_2 X_3 \cdots X_m) = P(X_1)P(X_2)P(X_3)\cdots P(X_m) \tag{2.18}$$

## Example 2.3

Assume that an engineering system has two critical subsystems $X_1$ and $X_2$. The failure of either subsystem can result in an accident or other safety-related problems. The probabilities of failure of subsystems $X_1$ and $X_2$ are 0.08 and 0.02, respectively. Calculate the probability of the occurrence of an accident or other safety-related problems if both of these subsystems fail independently.
  By inserting the given data values into Equation (2.17), we obtain

$$P(X_1 + X_2) = 1 - \prod_{i=1}^{2}(1 - P(X_i))$$

$$= P(X_1) + P(X_2) - [P(X_1)P(X_2)]$$

$$= 0.08 + 0.02 - \left[(0.08)(0.02)\right]$$

$$= 0.0984$$

Thus, the probability of the occurrence of an accident or other safety-related problems is 0.0984.

## 2.5   MATHEMATICAL DEFINITIONS

This section presents a number of mathematical definitions that will be useful in understanding subsequent chapters of this book.

### 2.5.1   PROBABILITY DENSITY FUNCTION

For a continuous random variable, the probability density function is expressed by [7]

$$f(t) = \frac{dF(t)}{dt} \qquad (2.19)$$

where
   $t$ is time (i.e., a continuous random variable).
   $f(t)$ is the probability density function.
   $F(t)$ is the cumulative distribution function.

### 2.5.2   CUMULATIVE DISTRIBUTION FUNCTION

For a continuous random variable, the cumulative distribution function is expressed by [7]

$$F(t) = \int_{-\infty}^{t} f(y)dy \qquad (2.20)$$

where
   $y$ is a continuous random variable.
   $F(y)$ is the cumulative distribution function.
   $f(y)$ is the probability density function.

For $t = \infty$, Equation (2.20) yields

$$F(\infty) = \int_{-\infty}^{\infty} f(y)dy$$

$$= 1 \qquad (2.21)$$

It simply means that the total area under the probability density curve is equal to unity.

Usually, in safety-related work, Equation (2.20) is simply written as

$$F(t) = \int_0^t f(y)dy \qquad (2.22)$$

### 2.5.3  EXPECTED VALUE

$$E(t) = \int_{-\infty}^{\infty} tf(t)dt \qquad (2.23)$$

where
  $E(t)$ is the expected value (i.e., mean value) of the continuous random variable $t$.

### 2.5.4  LAPLACE TRANSFORM

The Laplace transform (named after Pierre–Simon Laplace [1749–1827]) of a function, say $f(t)$, is expressed by

$$F(s) = \int_0^{\infty} f(t)e^{-st}dt \qquad (2.24)$$

where
  $s$ is the Laplace transform variable.
  $F(s)$ is the Laplace transform of $f(t)$.
  $t$ is a variable.

Laplace transforms of some commonly occurring functions in the engineering system safety-related studies are presented in Table 2.1 [9,10].

### 2.5.5  LAPLACE TRANSFORM: FINAL-VALUE THEOREM

If the following limits exist, then the final-value theorem may be expressed as:

$$\lim_{t \to \infty} f(t) = \lim_{s \to 0}\left[ sF(s)\right] \qquad (2.25)$$

### Example 2.4

Prove by using the following equation that the left side of Equation (2.25) is equal to its right side:

$$f(t) = \frac{\mu}{(\lambda + \mu)} + \frac{\lambda}{(\lambda + \mu)}e^{-(\lambda + \mu)t} \qquad (2.26)$$

where
  $\mu$ and $\lambda$ are constants.

**TABLE 2.1**
**Laplace Transforms of Some Commonly**
**Occurring Functions in Safety-Related Studies**

| No. | $f(t)$ | $F(s)$ |
|---|---|---|
| 1 | $T$ | $\dfrac{1}{s^2}$ |
| 2 | $t^n, n = 0,1,2,3,\ldots$ | $\dfrac{n!}{s^{n+1}}$ |
| 3 | $C$ (a constant) | $\dfrac{C}{s}$ |
| 4 | $e^{-\lambda t}$ | $\dfrac{1}{(s+\lambda)}$ |
| 5 | $te^{-\lambda t}$ | $\dfrac{1}{(s+\lambda)^2}$ |
| 6 | $\dfrac{df(t)}{dt}$ | $sF(s) - f(0)$ |
| 7 | $\lambda_1 f_1(t) + \lambda_2 f_2(t)$ | $\lambda_1 F_1(s) + \lambda_2 F_2(s)$ |

By substituting Equation (2.26) into the left side of Equation (2.25), we get

$$\lim_{t \to \infty}\left[ \frac{\mu}{(\lambda+\mu)} + \frac{\lambda}{(\lambda+\mu)} e^{-(\lambda+\mu)t} \right] = \frac{\mu}{(\lambda+\mu)} \tag{2.27}$$

With the aid of Table 2.1, the right side of Equation (2.25), and Equation (2.26), we get:

$$\lim_{s \to 0} s\left[ \frac{\mu}{s(\lambda+\mu)} + \frac{\lambda}{\lambda+\mu} \cdot \frac{1}{(s+\lambda+\mu)} \right] = \frac{\mu}{(\lambda+\mu)} \tag{2.28}$$

As the right sides of both Equations (2.27) and (2.28) are exactly the same, it proves that the left side of Equation (2.25) is equal to its right side.

## 2.6  PROBABILITY DISTRIBUTIONS

There are a large number of probability distributions in the published literature [11,12]. This section presents three probability distributions considered useful to perform engineering systems safety-related analysis.

### 2.6.1  EXPONENTIAL DISTRIBUTION

This is a widely used probability distribution to perform various types of reliability and safety studies concerning engineering systems. Its probability density function is defined by [13].

$$f(t) = \lambda e^{-\lambda t}, \text{ for } t \ge 0, \lambda > 0 \qquad (2.29)$$

where

$f(t)$ is the probability density function.

$t$ is time (i.e., a continuous random variable).

$\lambda$ is the distribution parameter (in the area of reliability/safety engineering, it is often referred to as unit/part/system failure rate).

By substituting Equation (2.29) into Equation (2.22), we obtain

$$F(t) = \int_0^t \lambda e^{-\lambda t} dt$$
$$= 1 - e^{-\lambda t} \qquad (2.30)$$

where $F(t)$ is the cumulative distribution function.

By inserting Equation (2.29) into Equation (2.23), we get

$$E(t) = \int_0^\infty t \lambda e^{-\lambda t} dt$$
$$= \frac{1}{\lambda} \qquad (2.31)$$

where $E(t)$ is the expected value or mean value of the exponential distribution.

## Example 2.5

Assume that the failure rate of an engineering system failing unsafely is 0.0005 failures per hour. Calculate the probability of the engineering system failing unsafely during a 800-hour mission by using Equation (2.30). Thus, in this case, we have $t = 800$ hours and $\lambda = 0.0005$ failures per hour.

By inserting these two values into Equation (2.30), we obtain

$$F(800) = 1 - e^{-(0.0005)(800)}$$
$$= 0.3296$$

Thus, the probability of the engineering system failing unsafely during the 800-hour mission is 0.3296.

## 2.6.2 RAYLEIGH DISTRIBUTION

This probability distribution is named after its founder, John Rayleigh (1842–1919) [1]. The distribution is frequently used in the theory of sound and from time to time in system reliability and safety-related studies as well. The distribution probability density function is defined by

$$f(t) = \left(\frac{2}{\theta^2}\right) t e^{-\left(\frac{t}{\theta}\right)^2}, \text{ for } t \geq 0, \theta > 0 \tag{2.32}$$

where

$\theta$ is the distribution parameter.

$t$ is time (i.e., a continuous random variable).

By substituting Equation (2.32) into Equation (2.22), we obtain the following equation for the cumulative distribution function:

$$F(t) = 1 - e^{-\left(\frac{t}{\theta}\right)^2} \tag{2.33}$$

By substituting Equation (2.32) into Equation (2.23), we get the following equation for the distribution expected value:

$$E(t) = \theta \Gamma\left(\frac{3}{2}\right) \tag{2.34}$$

where

$\Gamma(.)$ is the gamma function, which is defined by

$$\Gamma(m) = \int_0^\infty t^{m-1} e^{-t}, \text{ for } m > 0 \tag{2.35}$$

### 2.6.3 WEIBULL DISTRIBUTION

This probability distribution is named after its founder, W. Weibull, who developed it in the early 1950s [14]. The probability density function of the distribution is expressed by

$$f(t) = \frac{c t^{c-1}}{\theta^c} e^{-\left(\frac{t}{\theta}\right)^c}, \text{ for } t \geq 0, \theta > 0, c > 0 \tag{2.36}$$

where $c$ and $\theta$ are the shape and scale parameters, respectively.

By substituting Equation (2.36) into (2.22), we get the following equation for the cumulative distribution function:

$$F(t) = 1 - e^{-\left(\frac{t}{\theta}\right)^c} \tag{2.37}$$

Inserting Equation (2.36) into Equation (2.23), we get the following expression for the Weibull distribution expected value:

$$E(t) = \theta \Gamma\left(1 + \frac{1}{c}\right) \tag{2.38}$$

Finally, it is to be noted that for c = 1 and c = 2, the exponential and Rayleigh distributions are the special cases of this distribution, respectively.

## 2.7 SOLVING FIRST-ORDER DIFFERENTIAL EQUATIONS USING LAPLACE TRANSFORMS

Normally, Laplace transforms are used to find solutions to linear first-order differential equations in safety studies concerned with engineering systems. The following example demonstrates the finding of solutions to a set of linear-order differential equations concerned with engineering system safety.

### Example 2.6

Assume that an engineering system can be in any of the three states: operating normally, failed safely, or failed unsafely. The three first-order differential equations presented below describe the engineering system:

$$\frac{dP_0(t)}{dt} + (\lambda_s + \lambda_{us})P_0(t) = 0 \tag{2.39}$$

$$\frac{dP_1(t)}{dt} - \lambda_s P_0(t) = 0 \tag{2.40}$$

$$\frac{dP_2(t)}{dt} - \lambda_{us}P_0(t) = 0 \tag{2.41}$$

where
$P_i(t)$ is the probability that the engineering system is in state $i$ at time $t$, for $i$
    = 0 (operating normally), $i$ = 1 (failed safely), and $i$ = 2 (failed unsafely).
$\lambda_s$ is the engineering system constant safe failure rate.
$\lambda_{us}$ is the engineering system constant unsafe failure rate.

At time $t = 0$, $P_0(0) = 1$, $P_1(0) = 0$, and $P_2(0) = 0$.
Solve differential Equations (2.39)–(2.41) by using Laplace transforms.
By using Table 2.1, differential Equations (2.39)–(2.41), and the specified initial conditions, we get:

$$sP_0(s) - 1 + (\lambda_s + \lambda_{us})P_0(s) = 0 \tag{2.42}$$

$$sP_1(s) - \lambda_s P_0(s) = 0 \tag{2.43}$$

$$sP_2(s) - \lambda_{us}P_0(s) = 0 \tag{2.44}$$

Solving Equations (2.42)–(2.44), we obtain

$$P_0(s) = \frac{1}{(s + \lambda_s + \lambda_{us})} \tag{2.45}$$

$$P_1(s) = \frac{\lambda_s}{s(s + \lambda_s + \lambda_{us})} \tag{2.46}$$

$$P_2(s) = \frac{\lambda_{us}}{s(s + \lambda_s + \lambda_{us})} \tag{2.47}$$

The inverse Laplace transforms of Equations (2.45)–(2.47) are as follows:

$$P_0(t) = e^{-(\lambda_s + \lambda_{us})t} \tag{2.48}$$

$$P_1(t) = \frac{\lambda_s}{(\lambda_s + \lambda_{us})}\left[1 - e^{-(\lambda_s + \lambda_{us})t}\right] \tag{2.49}$$

$$P_2(t) = \frac{\lambda_{us}}{(\lambda_s + \lambda_{us})}\left[1 - e^{-(\lambda_s + \lambda_{us})t}\right] \tag{2.50}$$

Thus, Equations (2.48)–(2.50) are the solutions to differential Equations (2.39)–(2.41).

## 2.8   PROBLEMS

1. What is absorption law?
2. Mathematically define mean deviation.
3. Assume that the inspection department of an engineering system manufacturer inspected seven identical engineering systems and found 10, 4, 8, 15, 20, 30, and 10 defects in each engineering system. Calculate the average number of defects per engineering system.
4. Prove the Boolean algebra expression (2.12).
5. Define probability.
6. Assume that an engineering system has two critical subsystems $Y_1$ and $Y_2$. The failure of either subsystem can result in an accident or other safety-associated problems. The probabilities of failure of subsystems $Y_1$ and $Y_2$ are 0.06 and 0.07, respectively. Calculate the occurrence probability of an accident or other safety-associated problems if both of these subsystems fail independently.
7. Define the following two items:
   • Cumulative distribution function
   • Expected value
8. Define the following:
   • Laplace transform
   • Laplace transform: final-value theorem
9. Assume that the failure rate of an engineering system failing unsafely is 0.0002 failures per hour. Calculate the probability of the engineering system failing unsafely during a 400-hour mission.
10. What are the special-case distributions of the Weibull distribution?

# REFERENCES

1. Eves, H., *An Introduction to the History of Mathematics*, Holt, Rinehart and Winston, New York, 1976.
2. Owen, D.B., Ed., *On the History Statistics and Probability*, Marcel Dekker, New York, 1976.
3. Lipschutz, S., *Set Theory*, McGraw Hill Book Company, New York, 1975.
4. Speigel, M.R., *Statistics*, McGraw Hill Book Company, New York, 1961.
5. Speigel, M.R., *Probability and Statistics*, McGraw Hill Book Company, New York, 1975.
6. Nuclear Regulatory Commission, *Fault Tree Handbook*, Report No. NUREG-0492, U.S. Nuclear Regulatory Commission, Washington, D.C., 1981.
7. Mann, N.R., Schafer, R.E., Singpurwalla, N.D., *Methods for Statistical Analysis of Reliability and Life Data*, John Wiley and Sons, New York, 1974.
8. Lipschutz, S., *Probability*, McGraw Hill Book Company, New York, 1965.
9. Speigel, M.R., *Laplace Transforms*, McGraw Hill Book Company, New York, 1965.
10. Oberhettinger, E., Baddii, L., *Tables of Laplace Transforms*, Springer-Verlag, New York, 1973.
11. Patael, J.K., Kapadia, C.H., Owen, D.H., *Handbook of Statistical Distributions*, Marcel Dekker, New York, 1976.
12. Dhillon, B.S., *Reliability Engineering in Systems Design and Operation*, Van Nostrand Reinhold Company, New York, 1983.
13. Davis, D.J., An Analysis of Some Failure Data, *Journal of the American Statistical Association*, June 1952, pp. 113–150.
14. Weibull, W., A Statistical Distribution Function of Wide Applicability, *Journal of Applied Mechanics*, Vol. 18, 1951, pp. 293–297

# 3 Safety Basics

## 3.1 INTRODUCTION

Today safety has become a very important field, and its history in modern times may be traced back to 1868 when a patent was awarded for the first barrier safeguard in the United States [1]. In 1877, the Massachusetts legislature passed a law requiring appropriate safeguards on hazardous machinery, and in 1893, the United States Congress passed the Railway Safety Act [1,2].

Nowadays, the field of safety has developed into many areas including robot safety, software safety, medical equipment safety, transportation systems safety, and mining equipment safety. This chapter presents various introductory safety concepts considered quite useful to understand subsequent chapters of this book, which have been taken from published literature.

## 3.2 SAFETY AND ENGINEERS AND PRODUCT HAZARD CLASSIFICATIONS

The safety problem with engineering systems may be traced back to railroads. For example, a prominent English legislator was killed in a railroad accident the very day Stephenson's first railroad line was dedicated [2]. The following year, the boiler of the first locomotive built in the United States exploded and caused one death and badly injured a number of fuel servers [2,3]. Needless to say, modern engineering systems/products have become highly sophisticated and complex and their safety is a very challenging issue because of competition and other factors when all involved engineers are pressured to complete new designs rapidly as well as at a minimum cost. Past experiences over the years clearly indicate that this, in turn, generally led to more design-related deficiencies, errors, and causes of accidents.

Design-related deficiencies can cause or contribute to accidents, and they may take place because a designer/design [2,4]:

- Overlooked to foresee all unexpected applications of an item/product or its all potential consequences.
- Overlooked to eliminate or reduce human errors' occurrence.
- Overlooked to prescribe effective operational-related procedures in situations where hazards might exist.
- Overlooked to provide adequate protection in a user's personal protective equipment.
- Overlooked to warn properly of all potential hazards.

DOI: 10.1201/9781003212928-3

- Creates an arrangement of operating controls and other devices that increases reaction time in emergency conditions or is conducive to errors' occurrence.
- Incorporates weak warning mechanisms instead of providing a safe design for eradicating potential hazards.
- Does not appropriately consider or determine the action, error, omission, or failure consequences.
- Is confusing, incorrect, or unfinished.
- Places an unreasonable degree of stress on potential operators.
- Creates an unsafe characteristic of an item/product.
- Relies on product users for avoiding the occurrence of accidents.
- Violates general tendencies/capabilities of users.

There are many product hazards, and they may be grouped under the following six classifications [4,5]:

- **Classification I: Energy hazards**: These are of two types: potential energy hazards and kinetic energy hazards. The potential energy hazards pertain to parts such as electronic capacitors, compressed gas receivers that store energy, counterbalancing weights, and springs. During the equipment servicing process, such hazards are very important because stored energy can lead to serious injury when released suddenly.

    The kinetic energy hazards pertain to parts such as fan blades, flywheels, and loom shuttles because of their motion. Any object that interferes with the motion of such parts can experience substantial damage.
- **Classification II: Environmental hazards**: These are of two types: internal hazards and external hazards. The internal hazards are concerned with the changes in the surrounding environment that result in internally damaged products. This type of hazard can be minimized or eradicated altogether by considering factors such as electromagnetic radiation, illumination level, vibrations, atmospheric contaminants, extremes of temperatures, and ambient noise levels during the design process. The external hazards are the hazards posed by the product during its life span and include disposal hazards, maintenance-related hazards, and services-life operation hazards.
- **Classification III: Human factor hazards**: These are concerned with poor design in regard to humans, more specifically, to their physical strength, height, length of reach, weight, visual acuity education, visual angle, intelligence, computational ability, etc.
- **Classification IV: Electrical hazards**: These hazards have two principal elements: electrocution hazard and shock hazard. The major electrical hazard to product/property stems from electrical faults, often referred to as short circuits.
- **Classification V: Kinematic hazards**: These hazards are related to scenarios where parts come together while still moving and lead to possible cutting, crushing, or pinching of any object/item caught between them.

- **Classification VI: Misuse- and abuse-associated hazards**: These hazards are concerned with the products' usage by people. Past experiences over the years clearly indicate that the misuse of products can lead to serious injuries. Product abuse can also result in hazardous situations or injuries, and some of the causes for the abuse are lack of necessary maintenance and poor operating-related practices.

## 3.3  COMMON MECHANICAL-RELATED INJURIES AND PRODUCTS LIABILITY EXPOSURE'S COMMON CAUSES

In the day-to-day work environment in the industrial sector, people interact with various types of engineering equipment for performing tasks, such as cutting, punching, drilling, chipping, stamping, abrading, and shaping. There are various types of injuries that can take place during the performance of such tasks; some of the common ones are as follows [1]:

- **Breaking-associated injuries**: These injuries are generally concerned with machines used for deforming engineering materials. Often, a break in a bone is referred to as a fracture. In turn, fracture is grouped under many categories including transverse, complete, incomplete, comminuted, oblique, simple, and compound.
- **Crushing-associated injuries**: These injuries take place where a person's body part is caught between two hard surfaces moving progressively together and crushing any object/item that comes between them.
- **Shearing-associated injuries**: These injuries are concerned with shearing processes. In manufacturing, power-driven shears are commonly used for performing tasks such as severing metal, plastic, paper, and elastomers. In the past, from time to time during the use of such machines, tragedies, such as amputation of fingers/hands, have occurred.
- **Straining- and spraining-associated injuries**: For the occurrence of such injuries (e.g., straining of muscles or spraining of ligaments), in the industrial environment, there are numerous opportunities related to the use of machines or other tasks.
- **Puncturing-associated injuries**: These injuries take place when an object penetrates straight into a person's body and pulls straight out. In the industrial sector, generally, such injuries pertain to punching machines because they contain sharp tools.
- **Cutting- and tearing-associated injuries**: These injuries take place when a person's body part comes in contact with a sharp edge. The severity of a cut or tear depends upon the degree of damage to muscles, skin, arteries, veins, etc.

Past experiences over the years indicate that around 60% of the liability cases involved failure to provide appropriate danger warning labels on manufactured products/items. Nonetheless, some of the product liability exposure's common causes are as follows [6]:

- Poorly written instructions.
- Poorly written warnings.
- Faulty product design.
- Faulty manufacturing.
- Inadequate research during the product development process.
- Inadequate testing of product prototypes.

## 3.4 SAFETY MANAGEMENT PRINCIPLES, PRODUCT SAFETY MANAGEMENT PROGRAM, AND PRODUCT SAFETY ORGANIZATION TASKS

There are many safety management principles. The main ones are as follows [7,8]:

- Safety should be managed just like managing any other activity in an organization. More specifically, management should direct safety by setting attainable safety goals and by planning, organizing, and controlling to successfully attain the goals.
- Under most circumstances, unsafe behaviour is normal behaviour because it is the result of normal human beings reacting to their surrounding environment. Therefore, it is clearly the management's responsibility to conduct necessary changes to the environment that leads to unsafe behaviour.
- The safety system should be tailored to fit effectively the company culture.
- There are certain sets of circumstances that can be predicted to lead to severe injuries: high energy sources, non-productive activities, certain construction-related situations, and unusual, non-routine tasks.
- There is no single method/approach for achieving safety in an organization. But, for a safety system to be effective, it must satisfy certain criteria: have the top-level management visibly showing its support, be flexible, involve workers' participation, etc.
- The key to successful line safety performance is management-related procedures that clearly factor in accountability.
- The three important symptoms that highlight that something is not right in the management system are an unsafe act, an unsafe condition, and an accident.
- In building a good safety system, the three major subsystems that must be considered carefully are the managerial, the behavioural, and the physical.
- Safety's main function is to find and define the operational errors that result in accidents.
- The causes leading to unsafe behaviour can be highlighted, controlled, and classified.

The main objective of a product safety management program is to minimize an organization's exposure to product liability litigation and related problems. Thus, the key to minimizing liability exposure is to establish and effectively maintain a

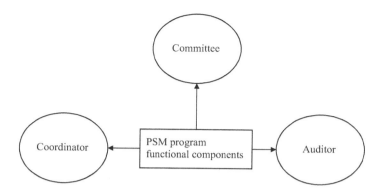

**FIGURE 3.1** Functional components of a product safety management (PSM) program.

comprehensive product safety management program. Generally, a product safety management (PSM) program has at least three functional components as shown in Figure 3.1 [1]. These are coordinator, committee, and auditor. The coordinator is concerned with coordinating and facilitating various units' involvement within the organization including the design, manufacturing, accounting, service, and marketing. The coordinator's level of authority plays a crucial factor in the success or failure of a PSM program.

Past experiences over the years clearly indicate that the higher the level of authority, the greater the chances the program will succeed. The National Safety Council (NSC) recommends that the PSM coordinator should have the authority for undertaking actions such as assisting in developing PSM program-associated policies; conducting PSM program audits; coordinating all program-associated documents; making necessary recommendations for product recalls, special analysis, product redesign, and field modifications; facilitating communication among all parties involved with the program, conduct complaint, incident, or accident analysis; establishing and maintaining relationships with organizations/agencies having product safety-associated missions, and developing a product safety information database for use by all parties involved in the program [1,9].

The committee is formed to deal with issues related to product safety. As the committee has representatives from all major units within the organization, it gives the coordinator of PSM a broad base of expertise to call upon and encourages broad-based support among all units. Finally, the auditor is concerned with evaluating the overall organization and units within it in regard to safety. Nonetheless, the auditor's specific duties include highlighting deficiencies in management commitment, observing the corrective measures put in place after the identification of product deficiencies, reviewing documentation of measures taken for rectifying product shortcomings, bringing deficiencies to management attention, and making appropriate corresponding recommendations [1].

A product safety organization conducts many tasks. The main ones are as follows [6,10]:

- Develop safety criteria on the basis of applicable voluntary and governmental standards for use by company design professionals, vendors, and subcontractors.
- Review all warning labels that are to be placed on the product in regard to compatibility to warnings in the instruction manuals, adequacy, and satisfying all legal requirements.
- Provide assistance to designers in choosing alternative means for eradicating or controlling hazards or other safety-associated problems in preliminary designs.
- Take part in reviewing accident-related claims or recall actions by government bodies and recommend appropriate remedial actions for justifiable claims or recalls.
- Determine if items, such as emergency equipment, protective equipment, or monitoring and warning devices, are really required for the product.
- Review all types of hazards and mishaps in current similar products for avoiding their repetition in the new products.
- Review the product to determine if all the potential hazards have been appropriately controlled or eradicated.
- Review all product test reports for determining deficiencies or trends in regard to safety.
- Review all proposed product operation and maintenance documents in regard to safety.
- Develop a system by which the safety program can be monitored effectively.
- Review governmental and non-governmental product safety-associated requirements.
- Review all safety-associated field reports and customer complaints.
- Prepare the product safety program and directives.

## 3.5   ACCIDENT CAUSATION THEORIES

There are many accident causation theories including the domino theory, the human factors theory, the epidemiological theory, the combination theory, the systems theory, and the accident/incident theory [1,6]. The first two of these theories are described below.

### 3.5.1   THE DOMINO THEORY

This theory is operationalized in ten statements called the "Axioms of Industrial Safety". These ten axioms were developed by H.W. Heinrich (American industrial safety pioneer), and they are as follows [1,6,11].

- **Axiom 1**: An accident can take place only when a person commits an unsafe act, and/or there is some physical or mechanical hazard.
- **Axiom 2**: Most accidents are due to the unsafe acts of humans.
- **Axiom 3**: The severity of an injury is largely fortuitous, and the specific accident that caused it is generally preventable.

- **Axiom 4**: Supervisors are the key individuals in the prevention of industrial accidents.
- **Axiom 5**: Management should assume safety responsibility with full vigour because it is in the best position for achieving final results.
- **Axiom 6**: The injuries' occurrence results from a completed sequence of events or factors, the final one of which is the accident itself.
- **Axiom 7**: The reasons why humans commit unsafe acts can be quite useful in selecting necessary corrective actions.
- **Axiom 8**: There are an accident's direct and indirect costs. Some examples of direct costs are medical-related costs, compensation, and liability claims.
- **Axiom 9**: An unsafe condition or an unsafe act by an individual does not always immediately result in an accident/injury.
- **Axiom 10**: The most effective accident prevention-related approaches are quite analogous with the quality and productivity methods.

Furthermore, as per Heinrich, there are five factors in the sequence of events leading up to an accident [1,6,11]:

- **Factor 1: Ancestry and social environment**: In this case, it is assumed that negative character traits, such as recklessness, avariciousness, and stubbornness that might lead individuals to behave unsafely, can be inherited through ancestry or acquired as a result of the social surroundings.
- **Factor 2: Fault of person**: In this case, it is assumed that negative character traits (i.e., whether inherited or acquired), such as violent temper, recklessness, nervousness, and ignorance of safety-related practices, constitute proximate reasons for committing unsafe acts or for the presence of physical or mechanical hazards.
- **Factor 3: Unsafe act/mechanical or physical hazard**: In this case, it is assumed that unsafe acts, such as starting machinery without warning, removing safeguards, and standing under suspended loads, committed by humans, and mechanical or physical hazards, such as inadequate light, unguarded gears, absence of rail guards, and unguarded point of operation, are the direct causes for the accidents' occurrence.
- **Factor 4: Accident**: In this case, it is assumed that events, such as falls of people and striking of people by flying objects, are examples of accidents that result in injury.
- **Factor 5: Injury**: In this case, it is assumed that the injuries directly caused by accidents include fractures and lacerations.

All in all, it is to be noted that the following two items are the central points of the Heinrich theory [6,12]:

(i) Injuries are the result of the action of all preceding factors.
(ii) The eradication of the central factor, i.e., unsafe act/hazardous conditions, definitely negates the action of preceding factors and, in turn, prevents the occurrence of accidents and injuries.

Additional information on the domino theory is available in Ref. [12].

## 3.5.2 THE HUMAN FACTORS THEORY

This theory is based on the assumption that accidents take place due to a chain of events caused by human error. There are three main factors that cause the occurrence of human errors [1,12]:

(i) Inappropriate activities.
(ii) Inappropriate response/incompatibility.
(iii) Overload.

The factor "inappropriate activities" is concerned with inappropriate activities conducted by a person due to human error. For example, a person misjudged the degree of risk involved in a stated task and then conducted the task on that misjudgement. The factor "inappropriate response/incompatibility" is another main human error causal factor, and three examples of inappropriate response by a person are as follows [6,12]:

### Example 1

A person totally disregarded the recommended safety-associated procedures.

### Example 2

A person totally removed a safeguard from a machine/equipment for improving output.

### Example 3

A person detected a hazardous condition but took no appropriate corrective action.

Finally, the factor "overload" is concerned with an imbalance between a person's capacity at any time and the load he/she is carrying in a given state. The capacity of a person is the product of a number of factors including the state of mind, fatigue, physical condition, stress, degree of training, and natural ability. The load carried by a person is composed of tasks for which he/she has responsibility, along with additional burdens due to the following three types of factors [6,12]:

- **Internal factors**: Three examples of these factors are personal problems, worry, and emotional stress.
- **Situational factors**: Two examples of these factors are unclear instructions and the level of risk.

- **Environmental factors**: Two examples of these factors are distractions and noise.

Additional information on the human factors theory is available in Ref. [12].

## 3.6  SAFETY PERFORMANCE MEASURING INDEXES

Over the years many indexes have been developed for measuring the safety performance of an organization. Two such indexes proposed by the American National Standards Institute are presented below [13].

### 3.6.1  DISABLING INJURY FREQUENCY RATE (DIFR)

This is defined by

$$DIFR = \frac{DI_t\left(1,000,000\right)}{EET} \tag{3.1}$$

where $DI_t$ is the total number of disabling injuries.

$EET$ is the employee exposure time expressed in hours.

The index is based on a total of four events that take place during the time period covered by the rate (i.e., temporary disabilities, permanent partial disabilities, permanent disabilities, and deaths).

One important benefit of this index is that it considers differences in the quantity of exposure due to varying worker/employee work hours, either within the framework of the organization during successive periods or among organizations categorized under a similar industry group [14].

### 3.6.2  DISABLING INJURY SEVERITY RATE (DISR)

This is defined by

$$DISR = \frac{D_{nc}\left(1,000,000\right)}{EET} \tag{3.2}$$

where $D_{nc}$ is the total number of days charged.

$EET$ is the employee exposure time expressed in hours.

This index is based on four factors occurring during the period covered by the rate (i.e., total scheduled charges (days) for all deaths, permanent total, and permanent partial disabilities, and the total number of days of disability from all temporary injuries). Some of the benefits of the index are as follows:

- Useful for making a meaningful comparison between different organizations.
- A quite useful tool to take into consideration differences in the quantity of exposure over time.

- A very useful tool to answer the question: "How serious are injuries in our organization?"

## 3.7 SAFETY COST ESTIMATION METHODS

Over the years several methods have been developed for estimating various types of safety-related costs. Three such methods are presented below.

### 3.7.1 THE SIMONDS METHOD

This method was developed by Professor R.H. Simonds of Michigan State College when working in conjunction with the National Safety Council [15]. Simonds reasoned that an accident's cost can be grouped into two main categories: insured and uninsured costs. Furthermore, Simonds stated that the insured cost can easily be estimated by simply examining some accounting-related records but the uninsured costs' estimation is more challenging. He proposed using the following relationship for estimating the total uninsured cost of accidents [1,15,16]:

$$TUC = \alpha_1 AC_1 + \alpha_2 AC_2 + \alpha_3 AC_3 + \alpha_4 AC_4 \tag{3.3}$$

where
  $TUC$ is the total uninsured cost of accidents.
  $\alpha_1$ is the total number of lost workday cases due to Class 1 accidents resulting in permanent partial disabilities and temporary total disabilities.
  $AC_1$ is the average uninsured cost associated with Class 1 accidents.
  $\alpha_2$ is the total number of physician's cases with Class 2 accidents, i.e., the occupational Safety and Health Act (OSHA) non-lost workday cases requiring treatment by a doctor.
  $AC_2$ is the average uninsured cost associated with Class 2 accidents.
  $\alpha_3$ is the total number of first-aid cases associated with Class 3 accidents, i.e., those accidents in which first aid was provided locally with a loss of working time of less than eight hours.
  $AC_3$ is the average uninsured cost associated with Class 3 accidents.
  $\alpha_4$ is the total number of non-injury cases associated with Class 4 accidents, i.e., those accidents causing minor injuries that do not require the attention of a medical professional.
  $AC_4$ is the average uninsured cost associated with Class 4 accidents.

### 3.7.2 THE HEINRICH METHOD

Over 90 years earlier, H.W. Heinrich categorically pointed out that for every single dollar of insured cost paid for accidents there were four dollars of uninsured costs borne by the company/organization [17]. All his conclusions were based on factors such as the review of 5,000 case files from organizations insured with a private company, research in the concerned organizations, and interviews with the staff members of the administrative and production services of all these enterprises [18]. Heinrich expressed "total cost of occupational-related injuries" as the sum of the direct and indirect costs. The cost is composed of the total benefits paid by the insurance company and the indirect cost, the expenditure assumed directly by the enterprise and is made up of the following elements:

- Lost time cost of workers who stop work are involved in the action.
- Lost time cost of first aid and hospital workers not paid by insurance.
- Cost of overheads for injured workers while in non-production mode.
- Injured workers' lost time cost.
- Cost to workers under welfare and benefits system.
- Cost related to profit and worker productivity loss.
- Management's lost time cost.
- Cost to workers in continuing wages of the insured.
- Machine/material damage-related cost.
- Lost orders' cost.
- Cost due to weakened morale.

### 3.7.3 THE WALLACH METHOD

This method was developed by M.B. Wallach in 1962 for analyzing the cost of the consequences of events concerning occupational-related injuries in various areas related to production: equipment, materials, manpower, machines, and time [19]. Although this approach highlights only the occupational-related injuries' effect on production, it has a distinct benefit of employing ideas and language quite familiar to organizations. Consequently, it is highly appealing to enterprises.

All in all, this method appears to be effective, particularly when measures are taken for quantifying occupational-related injuries' effect at the company level [18].

## 3.8   SAFETY COST PERFORMANCE MEASUREMENT INDEXES

There are many indexes that have been developed for measuring the overall safety cost performance of an organization. Unfortunately, as there is no single index adequate for determining an organization's overall safety cost effectiveness, several indexes in combination can be used for serving this purpose.

The true purpose of using these indexes is to indicate trends, utilizing the past safety cost performance as a point of reference, and encourage involved personnel for improving over the past experience. This section presents three safety cost performance associated indexes [15,18,20].

### 3.8.1   AVERAGE INJURY COST PER PROFIT DOLLAR INDEX

This index is used for determining the average cost of occupational-related injuries per profit dollar in an organization and is expressed by

$$AC_{oip} = \frac{TC_{oi}}{TP_d}$$ (3.4)

where

$AC_{oip}$ is the average cost of occupational-related injuries per profit dollar.
$TC_{oi}$ is the total cost of occupational-related injuries.
$TP_d$ is the total profit in dollars.

### 3.8.2 AVERAGE INJURY COST PER UNIT TURNOVER INDEX

This index is used for determining the average cost of occupational-related injuries per unit turnover in an organization and is expressed by

$$AC_{oiu} = \frac{TC_{oi}}{TU_t}$$ (3.5)

where

$AC_{oiu}$ is the average cost of occupational-related injuries per unit turnover.
$TC_{oi}$ is the total cost of occupational-related injuries.
$TU_t$ is the total number of units turnover (i.e., unit quantity, tons, etc.).

### 3.8.3 AVERAGE COST PER INJURY INDEX

This index is used for determining the average cost per occupational-related injury in an organization and is expressed by

$$AC_{oi} = \frac{TC_{oi}}{TOI}$$ (3.6)

where

$AC_{oi}$ is the average cost per occupational-related injury.
$TC_{oi}$ is the total cost of occupational-related injuries.
$TOI$ is the total number of occupational-related injuries.

## 3.9 PROBLEMS

1. Describe the following three types of product hazards:
   - Human factors hazards.
   - Energy hazards.
   - Kinematic hazards.
2. What are the commonly occurring mechanical-related injuries?
3. List at least five product liability exposure's common causes.
4. What are the main safety management principles?
5. Discuss functional components of a product safety management program.
6. What are the main tasks of a product safety organization?
7. Describe the human factors theory.

8. Discuss at least two indexes that can be used to measure safety cost performance in an organization.
9. Describe the Simonds safety cost estimation method.
10. Define the following three safety cost performance measurement indexes:
    * Average injury cost per profit dollar index.
    * Average injury cost per unit turnover index.
    * Average cost per injury index.

## REFERENCES

1. Goetsch, D.L., *Occupational Safety and Health*, Prentice Hall, Englewood Cliffs, NJ, 1996.
2. Hammer, W., Price, D., *Occupational Safety Management and Engineering*, Prentice Hall, Upper Saddle River, NJ, 2001.
3. Hunter, T.A., Operator Safety, Engineering, May 1974, pp. 358–363.
4. Dhillon, B.S., *Reliability, Quality, and Safety for Engineers*, CRC Press, Boca Raton, Florida, 2005.
5. Hunter, T.A., *Engineering Design for Safety*, McGraw Hill Book Company, New York, 1992.
6. Dhillon, B.S., *Engineering Safety: Fundamentals, Techniques, and Applications*, World Scientific Publishing, River Edge, NJ, 2003.
7. Petersen, D., *Techniques of Safety Management*, McGraw-Hill Book Company, New York, 1971.
8. Petersen, D., *Safety Management, American Society of Safety Engineers*, Des Plaines, IL, 1998.
9. NSC, *Accident Prevention Manual for Industrial Operations*, National Safety Council (NSC), Chicago, 1988.
10. Hammer, W., *Product Safety Management and Engineering*, Prentice Hall, Inc., Englewood Cliffs, NJ, 1980.
11. Heinrich, H.W., *Industrial Accident Prevention*, 4th ed., McGraw-Hill Book Company, New York, 1959.
12. Heinrich, H.W., Petersen, D., Roos, N., *Industrial Accident Prevention*, McGraw-Hill Book Company, New York, 1980.
13. Z-16.1, *Method of Recording and Measuring Work Injury Experience*, American National Standards Institute, New York, 1985.
14. Tarrants, W.E., *The Measuring of Safety Performance*, Garland STPM Press, New York, 1980.
15. Simonds, R.M., *Estimating Accident Cost in Industrial Plants*, Safety Practices Pamphlet No. III, National Safety Council, Chicago, 1950.
16. Raouf, A., Dhillon, B.S., *Safety Assessment: A Quantitative Approach*, Lewis Publishers, Boca Raton, Florida, 1994.
17. Heinrich, H.W., *Industrial Accident Prevention*, McGraw Hill Book Company, New York, 1931.
18. Andreoni, D., *The Cost of Occupational Accidents and Diseases*, International Labor Office, Geneva, Switzerland, 1986.
19. Wallach, M.B., Accident Costs: A New Concept, *Journal of the American Society of Safety Engineers*, July 1962, pp. 25–26.
20. Blake, R.P., Ed., *Industrial Safety*, Prentice Hall Inc., Englewood Cliffs, New Jersey, 1964.

5. Describe at least two systems that can be used to measure rotary core period motions in an organ, etc.

9. Describe the systematic survey and estimate a mat curve.

10. Using the following indices, list performance as percent indices:
   Average annual total dollar inflow index
   Average annual reserve inflow index
   Average consumption index

# REFERENCES

1. Ogrosky, H.O., Construction, Sedimentation, and Flood Control, Englewood Cliffs, N.J.

2. Linsley, R.K., et al., Applied Hydrology, McGraw-Hill, New York, 1982.

3. Clark, T.A., Organizational Behavior, McGraw-Hill, 1975, pp. 51-52.

4. Dunkin, J.S., Principles, Organizational Safety and Management, Prentice Hall, Boston, 2005.

5. Henry, T.A., Engineering Economics, Prentice-Hall, Englewood Cliffs, New York, 1982.

6. Dunham, H.S., Economics of Management, Performance and Operations, Waltham, Blaisdell Publishing, Reading, Mass., 2003.

7. Thuesen, H.G., Engineering Economy, McGraw-Hill Book Company, New York, 1971.

8. Grant, E.L., Principles of Engineering Economy, New Amsterdam, Des Plaines, Ill., 1960.

9. ASCE Manual of Engineering Standards for Subsurface Operations, National Subsoil Council, ASCE, Chicago, Ill.

10. Thuesen, H.G., et al., Engineering Economy, 5th ed., Prentice-Hall, Englewood Cliffs, N.J.

11. Grant, E.L., et al., Principles of Economic Analysis, McGraw-Hill Book Company, New York, 1982.

12. Newnan, D.W., et al., Principles of Engineering Economic Analysis, McGraw-Hill, New York, New York, 1980.

13. Riggs, J.L., et al., Engineering Economics, 3rd ed., McGraw-Hill, American Society for Engineering Education, New York.

14. Collins, W.D., The Economics of Engineering, 2nd edition, Canada, STAR Press, New York, 2000.

15. DeGarmo, E.M., et al., Engineering Economics in Non-structural and Quality Practices, Regulation, 10th edition, Prentice-Hall, 1954.

16. Collins, W.D., et al., Economics: A Course for Engineers, Lewis, New York.

17. Samuelson, P.W., Economics and Management, McGraw-Hill Book Company, New York.

18. Adelson, D., The Cost of Quality: Methodologies and Practices, Prentice-Hall Labor, Office Economics, Washington, D.C.

19. Walther, H.E., Problem Analysis and Quality Control, ASCE, American Society of Safety Engineers, Feb. 1989.

20. Taylor, F.W., et al., Industrial Engineering, McGraw-Hill, Inc., New York, 1984.

# 4 Safety Analysis Methods

## 4.1 INTRODUCTION

Although safety-related problems have been around for a very long time, the development of safety analysis methods is relatively new. Some of these methods were specifically developed for application in the safety area and others for application in different areas. Some examples of these methods are failure modes and effect analysis (FMEA), Markov method, quality control charts, and cause and effect diagram. FMEA and Markov methods were developed for use in reliability areas and quality control charts and the cause and effect diagram for application in quality control work. Nonetheless, the main objective of all these safety analysis methods is preventing the occurrence of accidents and hazards.

As the effectiveness of all these methods can vary quite significantly from one application to another, a careful consideration is necessary for selecting a method for a specific application. The safety analysis methods may be classified under two categories: mathematically based and non-mathematically based. For example, quality control charts and the Markov method belong to the mathematically based classification and FMEA and the cause-and-effect diagram to the non-mathematically based classification. This chapter presents a number of safety analysis methods considered very useful to perform safety analysis of engineering systems [1–6].

## 4.2 HAZARDS AND OPERABILITY ANALYSIS (HAZOP)

This method was developed for use in the chemical industry, and it is considered extremely useful for identifying safety-associated problems prior to the availability of full data concerning an item [7,8]. The fundamental objectives of the HAZOP are as follows [1–6]

- Produce a complete description of the process/facility.
- Review each process/facility element for determining how deviations from the design intentions can happen.
- Decide whether the deviations can lead to operating problems/hazards.

A HAZOP study is conducted by following the seven steps presented below [3,7,9]:

- **Step I**: Choose the system/process to be analyzed.
- **Step II**: Establish a team of appropriate experts.
- **Step III**: Describe the HAZOP process to all individuals forming the team.
- **Step IV**: Establish goals and time schedules.
- **Step V**: Carry out brainstorming sessions as considered appropriate.
- **Step VI**: Conduct analysis.
- **Step VII**: Document the study.

DOI: 10.1201/9781003212928-4

It is to be noted that HAZOP has basically the same weaknesses as FMEA. For example, both these methods predict problems related to system/process failure but do not factor human error into the equation. This is the key weakness because human error is often a factor in accidents.

## 4.3 FAILURE MODES AND EFFECT ANALYSIS

FMEA is a widely used method during the design process for analyzing engineering systems from their reliability aspect. It may simply be described as an effective approach for analyzing each potential failure mode in the system for examining the effects of such failure modes on the system [10]. The method can also be used to perform engineering systems safety analysis.

The history of FMEA goes back to the early years of the 1950s with the development of flight control systems, when the U.S. Navy's Bureau of Aeronautics, in order to develop a procedure for reliability control over the detailed design effort, developed a requirement called Failure Analysis [11]. Subsequently, the term "Failure Analysis" was changed to FMEA and in the 1970s, the U.S. Department of Defense directed its effort for developing a military standard titled "Procedures for Performing a Failure Mode, Effects, and Criticality Analysis" [12]. Basically, failure mode, effects, and criticality analysis (FMECA) is an extended version of FMEA. More specifically, when FMEA is extended for grouping each potential failure effect with respect to its severity (this includes documenting critical and catastrophic failures), the method is called FMECA [13].

The seven main steps followed for performing FMEA are as follows [14]:

- **Step I**: Define system boundaries and their associated detailed requirements.
- **Step II**: List all system components and subsystems.
- **Step III**: List each component's failure modes, the description, and the identification.
- **Step IV**: Assign probability/failure rate to each component failure mode.
- **Step V**: List each failure mode effect or effects on subsystems, the system, and the plant.
- **Step VI**: Enter remarks for each failure mode.
- **Step VII**: Review each critical failure mode and take appropriate action.

There are many factors that must be explored prior to the implementation of FMEA. Four of these factors are as follows [15,16]:

- **Factor I**: Careful examination of each and every conceivable failure mode by the involved professionals.
- **Factor II**: Obtaining engineer's approval and support.
- **Factor III**: Making decisions based on the risk priority number (RPN).
- **Factor IV**: Measuring FMEA cost/benefits.

Over the years, professionals working in the reliability analysis area have developed a number of guidelines/facts concerning FMEA. Some of these guidelines/facts are as follows [4,15]:

- Avoid developing the majority of the FMEA in a meeting.
- FMEA is not the tool for selecting the optimum design concept.
- FMEA is not designed for superseding the engineer's work.
- RPN could be misleading.
- FMEA has certain limitations.

Nonetheless, there are many advantages of conducting FMEA. Some of these are as follows [15,17]:

- A useful approach for improving communication among design interface personnel.
- A useful approach that starts from the detailed level and works upward.
- A systematic approach for classifying hardware failures.
- Provides safeguard against repeating the same mistakes in the future.
- Highlights safety concerns to be focused on.
- Serves as a useful tool for more efficient test planning.
- A visibility tool for management.
- Reduces development time and cost.
- A useful approach for comparing designs.
- Reduces engineering-related changes.
- Improves customer satisfaction.
- Easy to understand.

Additional information on this method is available in Ref. [17].

## 4.4   FAULT TREE ANALYSIS (FTA)

This is a widely used method for evaluating engineering systems from the reliability aspect during their design and development, particularly in the area of nuclear power generation. The method was developed in the early years of the 1960s at the Bell Telephone Laboratories for performing reliability analysis of the Minuteman Launch Control System [14,17]. A fault tree may simply be described as a logical representation of basic/primary events' relationship that leads to the occurrence of a specified undesirable event called the "top event" and is depicted using a tree structure with OR, AND, etc., logic gates.

Although there could be many purposes in performing fault tree analysis (FTA), some of the main ones are highlighting critical areas and cost-effective improvements, understanding the level of protection that the design concept provides against failures, understanding the functional relationship of the system failures, and satisfying jurisdictional requirements. The method can also be used for performing various types of safety analysis of engineering systems.

FTA starts by highlighting an undesirable event, called "top event", associated with a system. Fault events that can cause the top event's occurrence are connected and generated by logic operators, such as OR and AND. The OR gate provides a True output (i.e., fault) if one or more inputs (i.e., faults) are true. Similarly, the AND gate provides a True output (i.e., fault) if all the inputs (i.e., faults) are true.

A fault tree's construction proceeds by generating fault events in a successive manner until the fault events need not be developed any further. These fault events are known as primary events. During the fault tree construction process, one successively asks the question: "How could this fault event occur?".

Four basic symbols used in constructing fault trees are shown in Figure 4.1 [14,17].

All the symbols shown in Figure 4.1 are described below (it is to be noted that OR and AND gates are described again for the sake of clarity).

- **Circle**: It represents a basic fault event (e.g., failure of an elementary component). The event's parameters are the probability of occurrence, failure, and repair rates (the values of these parameters are generally obtained from empirical data).
- **Rectangle**: It represents a fault event that occurs from the logical combination of fault events through the input of a logic gate, such as OR and AND.
- **OR gate**: It denotes that an output event (fault) occurs if one or more of the input events (faults) occur.
- **AND gate**: It denotes that an output event (fault) occurs only if all of the input events (faults) occur.

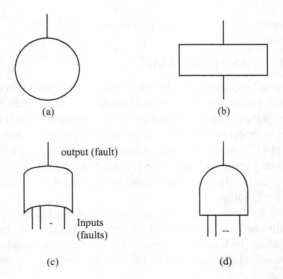

**FIGURE 4.1**    Fault tree basic symbols: (a) circle, (b) rectangle, (c) OR gate, (d) AND gate.

Information on other symbols used in conducting FTA is available in Refs. [14,17].

### Example 4.1

Assume that a windowless room has three light bulbs X, Y, and Z and one switch. The switch can fail to close. Develop a fault tree for the undesirable event (i.e., top fault event) "dark room" using the four symbols shown in Figure 4.1

In this case, the room can only be dark if there is no incoming electricity, all the three light bulbs burnt out, or the switch fails to close.

A fault tree for the example is shown in Figure 4.2.

## 4.4.1 FAULT TREE PROBABILITY EVALUATION

Under certain scenarios, it may be necessary for predicting the occurrence probability of a certain event (e.g., unsafe failure of an engineering system due to maintenance error). Before this could be achieved by utilizing the FTA method, the determination of the occurrence probability of output fault events of logic gates is needed.

Thus, the occurrence probability of the output fault event of an OR gate is given by [14,17]:

$$P_{or}(X) = 1 - \prod_{j=1}^{m}\left(1 - P(X_j)\right) \tag{4.1}$$

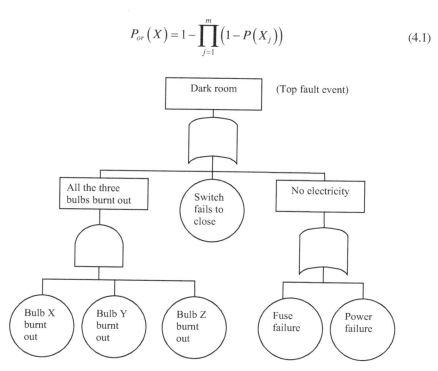

FIGURE 4.2  A fault tree for the top fault event: dark room.

where

$P_{or}(X)$ is the occurrence probability of OR gate's output fault event X.

$m$ is the number of input fault events.

$P(X_j)$ is the occurrence probability of input fault event $X_j$, for $j$=1,2,3,...,$m$.

Similarly, the occurrence probability of the output fault event of an AND gate is expressed by:

$$P_{an}(X) = \prod_{j=1}^{m} P(X_j) \tag{4.2}$$

where $P_{an}(X)$ is the occurrence probability of AND gate's output fault event X.

## Example 4.2

Assume that the probabilities of occurrence of fault events bulb $X$ burnt out, bulb $Y$ burnt out, bulb $Z$ burnt out, switch fails to close, fuse failure, and power failure shown in Figure 4.2 are 0.08, 0.07, 0.06, 0.05, 0.02, and 0.04, respectively. With the aid of Equations (4.1) and (4.2), calculate the occurrence probability of the top fault event "Dark room" and redraw the Figure 4.2 diagram with the calculated and specified data values.

By substituting the specified data values into Equation (4.2), the probability of occurrence of the event "All three bulbs burnt out" is

$$P_{an} = (0.08)(0.07)(0.06)$$

$$= 0.000336$$

where $P_{an}$ is the probability of occurrence of the event "All the three bulbs burnt out".

Similarly, by inserting the given data values into Equation (4.1), the probability of occurrence of the event "No electricity" is

$$P_{or} = 0.02 + 0.04 - (0.02)(0.04)$$

$$= 0.0592$$

where $P_{or}$ is the probability of occurrence of the event "No electricity".

By substituting the above-calculated values and the specified data value into Equation (4.1), the probability of occurrence of the top fault event "Dark room" is

$$P_{dr} = 1 - (1 - 0.000336)(1 - 0.05)(1 - 0.0592)$$

$$= 0.1065$$

Thus, the probability of occurrence of the top fault event "Dark room" is 0.1065. Figure 4.2 diagram with the calculated and the specified data values is shown in Figure 4.3.

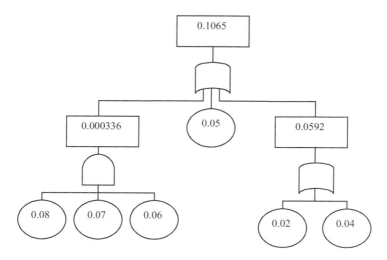

**FIGURE 4.3** A fault tree with the calculated and specified event occurrence probability values.

### 4.4.2 Fault Tree Analysis Advantages and Drawbacks

There are many advantages and drawbacks of the fault tree analysis. Some of its advantages are as follows [14,17]:

- Useful for providing insight into the system behaviour and to highlight failures deductively.
- Useful for handling complex systems more easily.
- Useful for providing options for management and others to conduct either quantitative or qualitative analysis.
- A graphic aid for management.
- Useful because it allows concentration on one specific failure at a time.
- Useful because it requires the analyst to understand thoroughly the system under consideration before starting the analysis.

In contrast, some of its drawbacks are as follows [14,17]:

- A time-consuming method.
- The end results are difficult to check.
- A costly approach.
- It considers parts in either working or failed state. More clearly, the partial failure states of the parts are difficult to handle.

Additional information on this method is available in Refs. [14,17].

## 4.5 MARKOV METHOD

This is a widely used method for performing various types of reliability analysis of engineering systems. The method is named after a Russian mathematician, Andrei A. Markov (1856–1922) and it can also be used for performing various types of safety analysis of engineering systems. The following three assumptions are associated with the Markov method [17,18]:

- All occurrences are independent of each other.
- The transitional probability from one system state to another in the finite time interval $\Delta t$ is given by $\lambda \Delta t$, where $\lambda$ is the transition rate (e.g., failure rate) from one system state to another.
- The probability of more than one transition occurrence in the finite time interval $\Delta t$ from one system state to another is very small or negligible (e.g., $(\lambda \Delta t)(\lambda \Delta t) \to 0$).

The following example demonstrates the application of the method.

### Example 4.3

Assume that an engineering system can either be in an operating or a failed state. The engineering system fails at a constant failure rate, $\lambda_s$, and is repaired at a constant repair rate, $\mu_s$. The system state space diagram is shown in Figure 4.4. The numerals in boxes denote system states. Obtain expressions for engineering system time-dependent and steady-state availabilities and unavailabilities, reliability, and mean time to failure by using the Markov method.

Using the Markov method, we write down the following equations for the engineering system states 0 and 1, respectively, shown in Figure 4.4:

$$P_0(t + \Delta t) = P_0(t)(1 - \lambda_s \Delta t) + P_1(t)\mu_s \Delta t \tag{4.3}$$

$$P_1(t + \Delta t) = P_1(t)(1 - \mu_s \Delta t) + P_0(t)\lambda_s \Delta t \tag{4.4}$$

where
   $t$ is time.
   $\lambda_s \Delta t$ is the probability of engineering system failure in finite time interval $\Delta t$.
   $(1 - \lambda_s \Delta t)$ is the probability of no failure in finite time interval $\Delta t$.

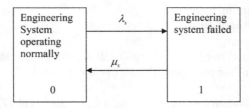

**FIGURE 4.4** Engineering system state space diagram.

$\mu_s\Delta t$ is the probability of engineering system repair in finite time interval $\Delta t$.
$(1-\mu_s\Delta t)$ is the probability of no repair in finite time interval $\Delta t$.
$P_0(t+\Delta t)$ is the probability of the engineering system being in operating state 0 at time $(t+\Delta t)$.
$P_1(t+\Delta t)$ is the probability of the engineering system being in failed state 1 at time $(t+\Delta t)$.
$P_i(t)$ is the probability that the engineering system is in state $i$ at time $t$, for $i = 0,1$.

From Equation (4.3), we obtain

$$P_0(t+\Delta t) = P_0(t) - P_0(t)\lambda_s\Delta t + P_1(t)\mu_s\Delta t \tag{4.5}$$

$$\lim_{\Delta t \to 0} \frac{P_0(t+\Delta t) - P_0(t)}{\Delta t} = -P_0(t)\lambda_s + P_1(t)\mu_s \tag{4.6}$$

Thus from Equation (4.6), we get:

$$\frac{dP_0(t)}{dt} + \lambda_s P_0(t) = \mu_s P_1(t) \tag{4.7}$$

Similarly, using Equation (4.4), we get:

$$\frac{dP_1(t)}{dt} + \mu_s P_1(t) = \lambda_s P_0(t) \tag{4.8}$$

At time $t = 0$, $P_0(0) = 1$ and $P_1(0) = 0$.
Solving Equations (4.7) and (4.8), we obtain

$$P_0(t) = \frac{\mu_s}{(\lambda_s+\mu_s)} + \frac{\lambda_s}{(\lambda_s+\mu_s)}e^{-(\lambda_s+\mu_s)t} \tag{4.9}$$

$$P_1(t) = \frac{\lambda_s}{(\lambda_s+\mu_s)} - \frac{\lambda_s}{(\lambda_s+\mu_s)}e^{-(\lambda_s+\mu_s)t} \tag{4.10}$$

Thus, the engineering system availability and unavailability, respectively, are

$$A_{es}(t) = P_0(t) = \frac{\mu_s}{(\lambda_s+\mu_s)} + \frac{\lambda_s}{(\lambda_s+\mu_s)}e^{-(\lambda_s+\mu_s)t} \tag{4.11}$$

and

$$UA_{es}(t) = P_1(t) = \frac{\lambda_s}{(\lambda_s+\mu_s)} - \frac{\lambda_s}{(\lambda_s+\mu_s)}e^{-(\lambda_s+\mu_s)t} \tag{4.12}$$

where
$A_{es}(t)$ is the engineering system time-dependent availability.
$UA_{es}(t)$ is the engineering time-dependent unavailability.

By letting time $t$ go to infinity in Equations (4.11) and (4.12), respectively, we obtain

$$A_{es} = \lim_{t \to \infty} A_{es}(t) = \frac{\mu_s}{(\lambda_s + \mu_s)} \tag{4.13}$$

and

$$UA_{es} = \lim_{t \to \infty} UA_{es}(t) = \frac{\lambda_s}{(\lambda_s + \mu_s)} \tag{4.14}$$

where
$A_{es}$ is the engineering system steady-state availability.
$UA_{es}$ is the engineering system steady-state unavailability.

By setting $\mu_s = 0$ in Equation (4.11), we get

$$R_{es}(t) = P_0(t) = e^{-\lambda_s t} \tag{4.15}$$

where $R_{es}(t)$ is the engineering system reliability at time $t$.

By integrating Equation (4.15) over the time interval $[0, \infty]$, we obtain the following equation for the engineering system mean time to failure [17]:

$$MTTF_{es} = \int_0^\infty e^{-\lambda_s t} dt$$
$$= \frac{1}{\lambda_s} \tag{4.16}$$

where $MTTF_{es}$ is the engineering system mean time to failure.

Thus, the engineering system time-dependent and steady-state availabilities are given by Equations (4.11) and (4.13), respectively. Similarly, its time-dependent and steady-state unavailabilities are given by Equations (4.12) and (4.14), respectively. Finally, the engineering system reliability and mean time to failure are given by Equations (4.15) and (4.16), respectively.

## Example 4.4

Assume that the engineering system constant failure rate and constant repair rates are 0.0005 failures/hour and 0.0008 repairs/hour, respectively. Calculate the engineering system availability during a 150-hour mission.

By substituting the given data values into Equation (4.11), we obtain

$$A_{es}(150) = \frac{0.0008}{(0.0005 + 0.0008)} + \frac{0.0005}{(0.0005 + 0.0008)} e^{-(0.0005 + 0.0008)(150)}$$

$$= 0.9318$$

Thus, the engineering system availability during a 150-hour mission is 0.9318.

## 4.6  CONTROL CHARTS

There are many types of control charts, and they were originally developed by Walter A. Shewhart in 1924 for use in the area of quality control [19]. They can also be used for performing various types of safety analysis. A control chart is a graphical technique used for evaluating whether a given process is in a "state of statistical control" or out of control. More clearly, when a sample value falls outside the upper and lower control limits of a control chart, it means that the process under consideration is out of statistical control and needs an investigation.

In safety-related work, the process could be the frequency of accidents, accidents' severity, etc. There are many types of control charts that can be utilized in safety-related studies. Here, their application to safety-associated problems is demonstrated through one type of control chart (i.e., the C-chart) only. The C-chart is based on the Poisson distribution. The Poisson distribution's mean and standard deviation are defined by [20]

$$\mu = \frac{A_t}{T_{tp}} \tag{4.17}$$

where
   $\mu$ is the Poisson distribution's mean.
   $A_t$ is the total number of accidents.
   $T_{tp}$ is the total time period.

and

$$\sigma = (\mu)^{1/2} \tag{4.18}$$

Where
   $\sigma$ is the Poisson distribution's standard deviation.
   Thus, the upper and lower control limits of the C-chart are defined by [21]

$$UCL_c = \mu + 3\sigma \tag{4.19}$$

and

$$LCL_c = \mu - 3\sigma \tag{4.20}$$

where
   $UCL_c$ is the upper control limit of the C-chart.
   $LCL_c$ is the lower control limit of the C-chart.

The following example demonstrates the C-chart's application to a safety-related problem.

**TABLE 4.1**
**Monthly Accident Occurrences**

| Month | No. of accidents |
|---|---|
| January | 20 |
| February | 10 |
| March | 5 |
| April | 15 |
| May | 10 |
| June | 20 |
| July | 8 |
| August | 12 |
| September | 6 |
| October | 4 |
| November | 16 |
| December | 6 |

**Example 4.5**

Assume that in an organization over a 12-month period, a total of 132 accidents occurred. Their monthly breakdowns are given in Table 4.1. Calculate the values of $\mu, \sigma$, and the upper and lower control limits of the C-chart and comment on the end results.

By substituting the given data into Equation (4.17), we get

$$\mu = \frac{132}{12} = 11 \text{ accidents/month}$$

Thus from Equation (4.18), we obtain

$$\sigma = (11)^{1/2} = 3.31$$

By inserting the above-calculated values into Equations (4.19) and (4.20), we get

$$UCL_c = 11 + 3(3.31) = 20.03$$

and

$$LCL_c = 11 - 3(3.31) = 1.07$$

Thus, the values of $\mu, \sigma$, and the upper and lower control limits of the C-chart are 11, 3.31, 20.93, and 1.07, respectively. It means that the occurrences of accidents are within the upper and lower control limits of the C-chart.

## 4.7 INTERFACE SAFETY ANALYSIS

Interface safety analysis (ISA) is concerned with determining the incompatibilities between assemblies and subsystems of an item/product that could lead to accidents.

The analysis establishes that totally distinct units/parts can be integrated into a quite viable system, and the normal operation of an individual unit or part will not deteriorate the performance or damage another unit/part or the whole system/product. Although ISA considers various relationships, they can be grouped basically under three classifications, as shown in Figure 4.5 [22].

The flow relationships may involve two or more items or units. For example, the flow between two items/units may entail lubricating oil, electrical energy, steam, fuel, water, or air. Furthermore, the flow also could be unconfined, such as heat radiation from one body to another. Generally, the common problems associated with many products are the proper flow of energy and fluids from one unit/item to another through confined spaces/passages, consequently leading to direct or indirect safety-associated problems/issues. Nonetheless, the flow-related problem causes include faulty connections between units and complete/partial interconnection failure. In the case of fluids, factors such as listed below must be considered carefully from the safety aspect [4,22].

- Loss pressure.
- Toxicity.
- Corrosiveness.
- Lubricity.
- Odor.
- Flammability.
- Contamination.

The physical relationships are connected to the units'/items' physical aspects. For example, two units/items might be well designed and manufactured and operate quite well individually, but they may fail to fit together due to dimensional-related differences or they may present other problems that may lead to safety-associated issues. Some typical examples of the other problems are as follows [4,22].

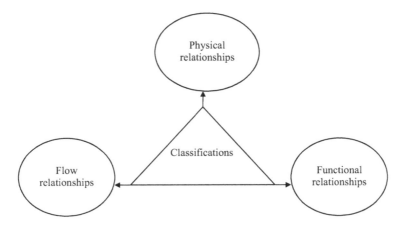

**FIGURE 4.5** Classifications of relationships considered by ISA.

- A very small clearance between units; thus the units may be damaged during the removal or replacement process.
- Impossible or restricted access to or egress from equipment.
- Impossible to join, tighten, or mate components/parts appropriately.

Finally, the functional relationships are concerned with multiple units/items. For example, in a circumstance where a unit's outputs constitute the inputs to a downstream unit, an error in outputs and inputs may directly or indirectly cause damage to the downstream unit, in turn, become a safety hazard. The condition of outputs could be:

- Zero outputs.
- Erratic outputs.
- Excessive outputs.
- Unprogrammed outputs.
- Degraded outputs.

## 4.8  PRELIMINARY HAZARD ANALYSIS (PHA)

PHA is relatively an unstructured method because of the unavailability of definitive information/data, such as functional flow diagrams and drawings. Thus, it is basically utilized during the concept design phase. Nonetheless, over the years, PHA has proved to be a very useful approach to take in the early steps of highlighting and eliminating hazards when the desirable data are unavailable. The findings of the PHA are extremely helpful for serving as a very useful guide in potential detailed analysis.

PHA requires the formation of an ad hoc team composed of members familiar with items such as material, equipment, substance, and/or process. The team members are asked to review hazard occurrences in the area of their expertise and, as a team, play the devil's advocate. Additional information on this method is available in Ref. [23].

## 4.9  ROOT CAUSE ANALYSIS (RCA)

RCA was developed by the U.S. Department of Energy for investigating industrial incidents, and it may simply be described as a systematic investigation approach that uses information collected during an assessment of an accident for determining the underlying factors deficiencies that caused the accident [24,25].

The ten general steps involved in conducting RCA are as follows [26,27]:

- **Step I**: Educate all individuals involved in RCA.
- **Step II**: Inform all appropriate staff members when a sentinel event is reported.
- **Step III**: Form an RCA team composed of appropriate personnel.
- **Step IV**: Prepare for and hold the first team meeting.

- **Step V**: Determine the event sequence.
- **Step VI**: Separate and highlight each event sequence that may have been directly or indirectly a contributory factor in the sentinel event occurrence.
- **Step VII**: Brainstorm about the factors surrounding the selected events that may directly or indirectly have been contributory to the sentinel event occurrence.
- **Step VIII**: Affinitize with the brainstorming session results.
- **Step IX**: Develop the appropriate action plan.
- **Step X**: Distribute the action plan and the RCA document to all individuals concerned.

Over the years, RCA has been applied in many areas and some of its benefits and drawbacks observed with its application are as follows [27,28]:

### 4.9.1 BENEFITS

- It is quite an effective tool for identifying and addressing systems and organizational-related issues.
- It is a well-structured and process-focused approach.
- The systematic application of RCA can uncover common root causes that link a disparate collection of accidents.

### 4.9.2 DRAWBACKS

- It is impossible for determining exactly if the root cause established by the analysis is really the actual cause for the accident's occurrence.
- It is quite impossible to be tainted by hindsight bias.
- It is a time-consuming and labour-intensive method.
- In essence, RCA is basically an uncontrolled case study.

## 4.10  TECHNIC OF OPERATION REVIEW (TOR)

TOR was developed in the early 1970s by D.A. Weaver of the American Society of Safety Engineers (ASSE) and seeks to highlight systemic causes rather than assigning blame in regard to safety [4,29,30]. This method allows management and employees to work jointly in conducting analysis of workplace-associated incidents, accidents, and failures, and it may be described simply as a hands-on analytical tool for highlighting the root system causes of an operation failure.

TOR uses a worksheet containing simple terms, basically, requiring "yes/no" decisions. An incident occurring at a certain point in time and location involving certain personnel activates TOR. Furthermore, it may be added that this method demands systematic evaluation of the actual circumstances surrounding the incident as it is not a hypothetical process.

The following eight steps are associated with this method [4,29,30]:

- **Step I**: Form the TOR team by selecting its members from all concerned areas.
- **Step II**: Hold a roundtable meeting for departing common knowledge to all team members.
- **Step III**: Highlight one key systemic factor that played an instrumental role in the incident's/accident's occurrence. This factor serves as a starting point for further investigation, and it must be based on the team consensus.
- **Step IV**: Use team consensus in responding to a sequence of "yes/no" options.
- **Step V**: Evaluate all the highlighted factors by ensuring the clear existence of team consensus with respect to the evaluation of each factor.
- **Step VI**: Prioritize all the contributory factors by starting with the most serious factor.
- **Step VII**: Develop appropriate preventive/corrective strategies in regard to each and every contributory factor.
- **Step VIII**: Implement all the strategies.

All in all, the main strength of this method is the involvement of line personnel in the analysis and its main weakness is an after-the-fact process.

## 4.11   JOB SAFETY ANALYSIS (JSA)

JSA is concerned with discovering and rectifying potential hazards that are intrinsic to or inherent in a given workplace.

Generally, individuals such as safety professionals, workers, and supervisors participate in JSA. The following five steps are associated with the performance of JSA [31]:

- **Step I:** Choose a job for analysis.
- **Step II:** Break down the job into various steps/tasks.
- **Step III**: Highlight all possible hazards and propose suitable measures for controlling them to appropriate levels.
- **Step IV**: Apply all the proposed measures.
- **Step V**: Evaluate the final results.

Past experiences over the years clearly indicate that the success of JSA depends on the degree of rigour exercised by the JSA team during the analysis process.

## 4.12   PROBLEMS

1. Describe hazards and operability analysis (HAZOP).
2. What are the advantages of performing failure modes and effect analysis (FMEA)?
3. Assume that a windowless room has two light bulbs A and B and one switch. The switch can fail to close. Develop a fault tree for the undesirable event (i.e., top fault event) "dark room".

4. Describe interface safety analysis.
5. Describe the following two items:
   - Preliminary hazard analysis.
   - Job safety analysis.
6. What are the advantages and disadvantages of the root cause analysis?
7. Assume that the engineering system constant failure rate and repair rate are 0.0008 failures/hour and 0.0009 repairs/hour, respectively. Calculate the engineering system availability during a 180-hour mission.
8. Compare FTA with FMEA.
9. Prove that the sum of Equations (4.11) and (4.12) is equal to unity.
10. What is the difference between FMEA and FMECA?

## REFERENCES

1. Roland, H.E., Moriarty, B., *System Safety Engineering and Management*, John Wiley and Sons, New York, 1983.
2. Gloss, D.S., Wardle, M.G., *Introduction to Safety Engineering*, John Wiley and Sons, New York, 1984.
3. Goetsch, D.L., *Occupational Safety and Health*, Prentice Hall, Englewood Cliffs, NJ, 1996.
4. Dhillon, B.S., *Engineering Safety: Fundamentals, Techniques, and Applications*, World Scientific Publishing, River Edge, NJ, 2003.
5. Raouf, A., Dhillon, B.S., *Safety Assessment: A Quantitative Approach*, Lewis Publishers, Boca Raton, FL, 1994.
6. Dhillon, B.S., *Reliability, Quality, and Safety for Engineers*, CRC Press, Boca Raton, Florida, 2005.
7. American Institute of Chemical Engineers, *Guidelines for Hazard Evaluation Procedures*, American Institute of Chemical Engineers, New York, 1985.
8. Dhillon, B.S., Rayapati, S.N., Chemical Systems Reliability: A Survey, *IEEE Transactions on Reliability*, Vol. 37, 1988, pp. 199–208.
9. Canadian Standards Association, *Risk Analysis Requirements and Guidelines*, Report No. CAN/CSA-Q634-91, prepared by the Canadian Standards Association, 1991. Available from Canadian Standards Association, 178 Rexdale Boulevard, Rexdale, Ontario, Canada.
10. Omdahl, T.P., Ed., *Reliability, Availability, and Maintainability (RAM) Dictionary*, American Society for Quality Control (ASQC) Press, Milwaukee, WI, 1988.
11. MIL-F-18372 (Aer), *General Specification for Design, Installation, and Test of Aircraft Flight Control Systems*, Bureau of Naval Weapons, U.S. Department of the Navy, Washington, D.C.
12. MIL-STD-1629, *Procedures for Performing a Failure Mode, Effects, and Criticality Analysis*, U.S. Department of Defense, Washington, D.C., 1980.
13. Jordan, W.E., Failure Modes, Effects, and Criticality Analysis, Proceedings of the Annual Reliability and Maintainability Symposium, 1972, pp. 30–37.
14. Dhillon, B.S., Singh, C., *Engineering Reliability: New Techniques and Applications*, John Wiley and Sons, New York, 1981.
15. Palady, P., *Failure Modes and Effects Analysis*, PT Publications, West Palm Beach, Florida, 1995.
16. McDermott, R.E., Mikulak, R.J., Beauregard, M.R., *The Basic of FMEA*, Quality Resources, New York, 1996.

17. Dhillon, B.S., *Design Reliability: Fundamentals and Applications*, CRC Press, Boca Raton, Florida, 1999.
18. Shooman, M.L., *Probabilistic Reliability: An Engineering Approach*, McGraw-Hill Book Company, New York, 1968.
19. *Statistical Quality Control Handbook*, AT and T Technologies, Indianapolis, IN, 1956.
20. Ireson, W.G., Ed., *Reliability Handbook*, McGraw-Hill, New York, 1966.
21. Jourdan, W.E., Failure Modes, Effects and Criticality Analysis, Proceedings of the Annual Reliability and Maintainability Symposium, 1972, pp. 30–37.
22. Hammer, W., *Product Safety Management and Engineering*, Prentice Hall, Inc., NJ, 1980.
23. Bahr, N.J., *System Safety Engineering and Risk Assessment: A Practical Approach*, CRC Press, Boca Raton, FL, 1997.
24. Latino, R.J., *Automating Root Cause Analysis, in Error Reduction in Health Care*, edited by P.L. Spath, John Wiley and Sons, New York, 2000, pp. 155–164.
25. Busse, D.K., Wright, D.J., *Classification and Analysis of Incidents in Complex, Medical Environments*, Report, 2000. Available from the Intensive Care Unit, Western General Hospital, Edinburgh, U.K.
26. Burke, A., *Root Cause Analysis*, Report, 2000. Available from the Wild Iris Medical Education, P.O. Box 257, Comptche, CA.
27. Dhillon, B.S., *Human Reliability and Error in Medical System*, World Scientific Publishing, River Edge, NJ, 2003.
28. Wald, H., Shojania, K.G., Root Cause Analysis, in *Making Health Care Safer: A Critical Analysis of Patient Safety Practices*, edited by A.J. Markowitz, Report No. 43, Agency for Health Care Research and Quality, U.S. Department of Health and Human Services, Rockville, MD, 2001, Chapter 5, pp. 1–7.
29. Hallock, R.G., Technique of Operations Review Analysis: Determine Cause of Accident/Incident, *Safety and Health*, Vol. 60, No. 8, 1991, pp. 38–39.
30. Goetsch, D.L., *Occupational Safety and Health*, Prentice Hall, Englewood Cliffs, NJ, 1996.
31. Hammer, W., Price, D., *Occupational Safety Management and Engineering*, Prentice Hall, Englewood Cliffs, NJ, 2001.

# 5 Robot Safety

## 5.1 INTRODUCTION

Nowadays, robots are being used in many diverse areas and applications, and over the years their safety-related problems have increased quite significantly. Each area and application may call for specific precautions for operators, maintenance workers, programmers, robot systems, and so on. In order to meet needs such as these, over the years various organizations around the world have developed documents that are specifically concerned with robot safety.

Two examples of such documents are the Japanese Industrial Safety and Health Association document titled "An Interpretation of the Technical Guidance on Safety Standards in the Use, Etc., of Industrial Robots" [1] and the American National Standard for Industrial Robots and Robot Systems-Safety Requirements [2].

Usually, factors such as increasing productivity and replacing humans in conducting hazardous and difficult tasks play an important role in the use of robots in the industrial sector. It is to be noted that improper consideration of safety in robotics planning may generate hazardous conditions other than those that the robot may be replacing. It simply means that for successful applications of robots, during the planning phase safety must be considered with utmost care.

This chapter presents various important aspects of robot safety.

## 5.2 ROBOT SAFETY-RELATED PROBLEMS AND HAZARDS

There are many unique robot safety-related problems faced by safety professionals concerned with robots. Nine of these problems are as follows [3,4,5]:

- **Problem I**: A robot may lead to a high risk of fire or an explosion if it is placed in an unfriendly environment.
- **Problem II**: Robot mechanical design-associated problems may result in hazards such as pinching, grabbing, and pinning.
- **Problem III**: Robots generate potentially hazardous conditions because they manipulate items of varying sizes and weights.
- **Problem IV**: The presence of a robot receives great attention from humans in the surrounding area, who are often quite ignorant of the possible associated hazards.
- **Problem V**: Various safety-associated electrical design problems can occur in robots. Some examples of these problems are potential for electric shock, poorly designed power sources, and fire hazards.
- **Problem VI**: Robot-associated maintenance procedures may lead to hazardous situations.

DOI: 10.1201/9781003212928-5

- **Problem VII**: Generally, robots function quite closely with other machines and humans in the same environment. In particular, humans are subject to collisions with the robots' moving parts, tripping over loose control/power cables (if any), and being pinned down.
- **Problem VIII**: Management attitudes very much lead to miscomprehension of robot safety-associated concepts.
- **Problem IX**: In the event of the occurrence of a control, hydraulic, or mechanical failure, robots may move out of their programmed area zones and strike something or they may throw some small part/item.

There are three basic types of robot hazards as shown in Figure 5.1 [3–6]. These are impact hazard, trapping hazard, and the hazards that develop from the application.

The impact hazard involves being struck by a robot's moving part or by parts/items being carried by the robot. It is to be noted that this hazard also includes being struck by flying objects that are ejected or dropped by the robot. The trapping hazard is generally the result of robot movements in regard to fixed objects such as machines and pests in the same area. The auxiliary equipment's movements could be the other possible causes. Two examples of such equipment are carriages and pallets.

Some of the hazards that develop from the application are the exposure to toxic substances, burns, arc flash, and electric shocks. The most prevalent causes of these hazards are human error, control errors, unauthorized access, and mechanical-related problems [7]. Human error occurs when humans enter the robot's protected work zone and conduct their required tasks close to it. Generally, complacency or lack of care during human interaction leads to human error. Control errors may simply be expressed as intrinsic faults within the robot's control system. Some examples of these errors are software problems, electrical interference, and faults in the hydraulic, pneumatic, or electrical subcontrol systems.

Unauthorized access is probably the most prevalent source of hazard; it can be easily controlled by developing appropriate standards for access and then following these effectively. Finally, mechanical-related problems arise from the application or

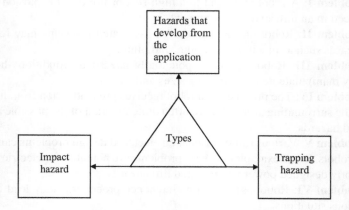

**FIGURE 5.1**    Basic types of robot hazards.

the robot system itself. Normally, mechanical hazards take place during the transfer of items with sharp edges, and from malfunctions due to poor maintenance or overloading.

## 5.3  ROBOT MANUFACTURERS' AND USERS' ROLES IN ROBOT SAFETY

As both manufacturers and users of robots profit from robot-related safety, each has a very important role to play in its effective promotion. Robot manufacturers' basic responsibility is to design and manufacture safe robots. The design-related safety measures of robots can be grouped under the following five categories [4,7–9]:

- **Category I: Electrical/Electronic subsystem.** There are various safety-associated features that can be designed into the robot's electrical/electronic subsystem including intrinsic safety circuits, emergency stops that can be activated by an operator in the event of a hazardous situation, equipping the teach pendant with a dead-man switch, and input/output signal conditioning.
- **Category II: Mechanical subsystem.** Safety in mechanical subsystems of robots can be built in the design process, by taking into consideration the sources of hazards and appropriate measures for eliminating them. In the mechanical subsystem, the main sources of hazard are poor reliability, incompatibility of materials with the work environment, kinetic energy storage capacity, and pinch points. Furthermore, during the design process, proper attention must be paid to specific sources of danger such as exposed motors, flopping cables and hoses, uncovered ball screws, and protruding linkages.
- **Category III: Operational and control subsystems' software.** The most safety-associated awareness related to the robot can be introduced in the software design for the operational and control subsystem. However, it is to be noted that the capabilities that can be designed into robots are only useful if the logic system is active as well as the robots themselves are calibrated appropriately. Some examples of the capabilities are as follows:
  - A response to interrupts generated by limits associated with hardware travel.
  - A response system to communication-related data-flow abnormalities.
  - Imposition of a maximum robot-operating speed during teaching.
  - An emergency response system to an unusual velocity on each and every server-controlled drive.
  - A response system to switch-off signals from outside interfaces.
- **Category IV: Control subsystem's algorithm.** Robot-associated safety can be improved appreciably through the effective control subsystem algorithm design. Therefore, proper attention has to be paid to items such as the regulation and control of the drives' maximum operating speed and the definition of the tolerable feedback system's following error.

- **Category V: Operational procedures**. These procedures are robot design's integral part, and they must be provided effectively for the safe operation of the robot. They include items such as appropriate precautions, operating manuals, and sets of instructions.

Past experiences over the years clearly indicate that robot users play an equally important role with regard to robot-related safety. The robot users' safety-related responsibilities may be grouped under the following three categories [9]:

- **Category I: The management**. In regard to safety, the basic two responsibilities of the robot user's management team are in appreciating the robotization's safety implications as well as supporting the safety department/unit to promote the special features of robot safety.
- **Category II: The safety department**. The safety department has many responsibilities, including installing appropriate barriers with interlocking gates at the work boundary, keeping abreast with the latest developments associated with robot-related safety, installing appropriate warning signs or other appropriate measures in the robot work envelope, collaborating with other groups concerned with the purchase and the use of robots, and providing appropriate safety-related training to all concerned personnel.
- **Category III: The engineering/maintenance department**. The responsibilities of the engineering/maintenance department include the installation according to the manufacturer's instructions, a proper maintenance program, effective training of maintenance workers, and locating all control stations, with the exception of the Pendant Control (whenever required) outside the restricted operating zone.

## 5.4  SAFETY-RELATED CONSIDERATIONS IN ROBOT DESIGN, INSTALLATION, PROGRAMMING, AND OPERATION AND MAINTENANCE PHASES

In order to minimize robot safety-related problems, it is absolutely essential to properly consider the safety factor during the robot design, installation, programming, and operation and maintenance phases. Some safety-related guidelines considered very useful concerning each of these four phases are presented below, separately [1,10–12].

### 5.4.1  ROBOT DESIGN PHASE

The robot design safety-related features may be categorized under the following three classifications:

- **Classification I: Electrical**. The electrical safety features include designing wire circuitry capable of stopping the robot's movement and locking its brakes, eliminating the risk of an electric shock, ensuring the internal safety of the robot so that it will not ignite in a combustible environment, minimizing the effects of electromagnetic and radio frequency interferences, having a built-in hose and cable routes using adequate insulation, having a fuse "blow" long before human crushing pressure is experienced, and sectioning and providing panel covers.

- **Classification II: Mechanical**. The mechanical safety features include designing teach pendant ergonomically, having drive mechanism covers, eliminating sharp corners, having several emergency stop buttons, putting guards on items such as gears, pulleys, and belts, ensuring the existence of mechanisms for releasing the stopped energy, and having dynamic brakes for software crash or power failure.
- **Classification III: Software**. The software safety features include periodically examining the built-in self-checking software for safety, providing a robot motion simulator, using a procedure of checks to determine why a failure occurred, prohibiting a restart by merely resetting a single switch, having built-in commands, having a restart approach after experiencing an emergency stop, and having a standby power source for the robot's functioning with programs in random access memory.

### 5.4.2 ROBOT INSTALLATION PHASE

There are many robot installation phase-related safety features. Some of these features are presented below.

- Providing an appropriate illumination level to humans concerned with the robot.
- Using vibration-reducing pads when necessary.
- Distancing all circuit boards from electromagnetic fields.
- Highlighting the danger zones with the aid of codes, signs, line markings, and so on.
- Installing all electrical cables according to electrical codes.
- Placing an appropriate shield between humans and the robot.
- Controlling all environmental factors as necessary.
- Providing protection to control circuitry by filtering surges and spikes.
- Placing all robot controls outside the hazard area.
- Installing the required interlocks for interrupting robot motion.
- Installing appropriate sensing devices, interlocks, and so on.
- Adding cushions, pads, and so on to all possible collision points with humans.
- Ensuring the accessibility and visibility of emergency stops.
- Labelling all the stored energy sources.

### 5.4.3 ROBOT PROGRAMMING PHASE

Safety during the robot programming phase is as important as during its previous two phases (i.e., design and installation phases). A study of 131 cases reported that setters/programmers were at the highest risk, accounting for 57% of the accidents;

the comparative figures for fault clearance personnel, operators, and maintenance personnel (servicing/repair) were 26%, 13%, and 4%, respectively [12].

There are various factors that, directly or indirectly, characterize a robot programmer's work conditions, including being subject to stress by improper lighting, working in a bending position (i.e., generally in the robot's movement zone), and frequently changing position because the torch's tip is concealed by the clamping device. Some of the safety-related measures that can be taken into consideration with respect to robot-related safety are as follows [12]:

- Designing the programmer's work area in such a way that it eliminates unnecessary stress.
- Turning off safety-associated devices with a key only.
- Marking all the programming positions.
- Pressure-sensitive mats on the floor at the programmer's position.
- Locked turntables during programming.
- Mandatory reduced speed.
- Planning the programmer's location outside the movement zone (i.e., in a space facing the robot, semiraised on a platform, etc.).
- A manual programming device that contains an emergency off switch.
- Hold-to-run-buttons.

### 5.4.4 Robot Operation and Maintenance Phase

There are many safety-related measures associated with the robot operation and maintenance phase [3,8,10]. Some of these measures require actions such as follows:

- Ensure that only authorized and trained personnel operate and maintain robots, report, investigate, and repair any faults or unusual robot motions promptly.
- Develop the necessary safety operations and maintenance procedures.
- Provide the initial and periodic training to people associated with robot operation and maintenance.
- Ensure the operational readiness of all safety devices (i.e., guards, barriers, and interlocks).
- Perform preventive maintenance regularly and use only the approved parts.
- Make certain that all emergency stops are functional.
- Observe all government codes and other regulations concerned with the operation and maintenance of robots.
- Block out all concerned power sources during maintenance.

## 5.5 ROBOT SAFETY-RELATED PROBLEMS CAUSING WEAK POINTS IN PLANNING, DESIGN, AND OPERATION

Past experiences over the years clearly indicate that there are many weak points in planning, design, and operation, which result in robot safety-related problems in an industrial setting [2,7]. Some of the weak points associated with planning are as follows:

- **Improper safety devices**. These comprise improper guards (i.e., containing gaps, being close to hazard points, or too low) and faulty emergency shutdown circuits.
- **Unsafe or confused linkages**. These linkages are basically concerned with interfaces between individual machines.
- **Poor work organization**. This is a very important factor, particularly in programming and stoppages.
- **Poor spatial arrangement**. This can lead to confusion and the possibility of collision.

Similarly, some of the weak points in design in regard to robot safety in the industrial sector are shown in Figure 5.2 [13].

Finally, some of the weak points in operational procedure in regard to robot safety in industrial settings are as follows:

- Failure to provide feedback to all personnel involved in design and layout concerning weak spots and how to ensure their removal appropriately.
- Poor training to all workers who are directly or indirectly involved with industrial robots.
- Allowing counter-safety working procedures during a stoppage.

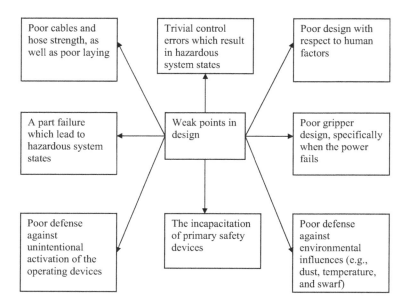

**FIGURE 5.2**    Weak points in design in regard to robot safety.

## 5.6 COMMON ROBOT SAFETY-RELATED FEATURES AND THEIR FUNCTIONS

Generally, a well-designed robot incorporates, to the greatest extent possible, safety-related features that clearly take into consideration all modes of robot operation (i.e., maintenance, programming, and normal operation) [4,14,15]. Nonetheless, some of the safety-related features are usually common to all robots and the others are specific to the robot types. Some of the common robot safety-related features, along with their corresponding intended functions, given in parentheses, are presented below [4,14].

- Slow-speed control (it allows program execution at reduced speeds).
- Power-on button (it energizes all machine power).
- Hardware stops (absolute control on movement/travel limits).
- Power disconnect switch (it removes all power at the machine junction box).
- Line indicator (it highlights that incoming power is connected at the junction box).
- Stop button (it removes manipulator and control power).
- Automatic/manual dump (it provides means for relieving pneumatic/hydraulic pressure).
- Software stops (computer-controlled travel limit).
- Condition indicators and messages (these provide a visual indication by lights or display screens of the system condition).
- Arm-power-only button (it applies power to the manipulator only).
- Program reset (this drops the system out of playback mode).
- Servo-motor brake (it maintains the arm position at a standstill).
- Teach pendant trigger (it must be held by the operator for arm power in teach mode).
- Remote connections (these allow remote control of essential machine/safety functions).
- Hydraulic fuse (it protects against high-speed force/motion in teach mode).
- Teach/playback mode selector (it provides the involved operator with control over the operating mode).
- Control-power-only button (it applies power to the control section only).
- Parity checks, error detecting, and so on (whereby the computer approaches for self-checking a variety of functions).
- Hold/run button (it stops arm motion, but leaves power on).
- Step button (it allows program execution one step at a time).

## 5.7 ROBOT SAFEGUARD APPROACHES

There are many robot safeguard approaches [6]. Four of these approaches are described below, separately [4,6,7].

### 5.7.1 PHYSICAL BARRIERS

These barriers are considered quite useful for safeguarding humans in the work areas of robots. However, in many cases, they are not the ultimate solution to a robot safety-related problem. The objective of such barriers is stopping humans from reaching over, around, under, or through the barriers into the prohibited robot work zone [2]. Some examples of physical barriers are plastic safety chains, chain-link fences, tagged-rope barriers, and safety rails.

Some guidelines considered quite useful concerning physical barriers are as follows [7]:

- Avoid trapping points within a barrier's framework by providing a sufficient buffer space between the barrier and the work area.
- Make use of fences in situations where long-range projectiles are considered a hazard.
- Use safety rails in circumstances where projectiles are clearly not a problem at all.
- Chain-link fences and safety rails are very useful in situations where intrusion is considered a particular problem.

In any case, when the installation of a peripheral physical barrier is considered, the following five questions have to be asked clearly [16]:

(i) Are the perimeter dimensions reliable?
(ii) How were perimeter dimensions established?
(iii) What is being protected?
(iv) Is the protection effective?
(v) Can the perimeter be bypassed easily?

### 5.7.2 FLASHING LIGHTS

This robot safeguard approach calls for the installation of a flashing light on the robot itself or at the perimeter of the robot work area. The flashing light's objective is to alert people in the area that programmed motion is happening or could happen at any time.

It is to be noted that when such an approach (i.e., flashing lights) is employed, ensure that the flashing light is energized continuously during the period when the robot drive power is activated.

### 5.7.3 WARNING SIGNS

The warning signs are employed in situations where robots, by virtue of their speed, size, and inability for imparting a significant amount of force, cannot injure people. Examples of such conditions are laboratory and small-part assembly robots. Robots such as these need no special safeguarding as warning signs are sufficient for the uninformed individuals in the area.

However, it is to be noted that warning signs are useful for all robot application areas, irrespective of whether robots possess the ability for injuring humans or not.

### 5.7.4 INTELLIGENT SYSTEMS

These systems make their decisions through remote sensing, software, and hardware. In order to achieve an effective intelligent collision-avoidance system, the robot's operating environment has to be restricted properly and special sensors and software have to be employed widely. This calls for the need for a highly sophisticated computer for making the right decisions and real-time computations.

Finally, it is to be noted that in most industrial settings, it is usually not possible to restrict the environment.

## 5.8   ROBOT WELDING OPERATIONS-ASSOCIATED SAFETY CONSIDERATIONS

Nowadays, in the industrial sector, robots are used for performing various types of welding operations. Welding robots undertake welding processes such as laser, gas metal arc, gas tungsten arc, and resistance (spot) [7,17].

There are many factors that have to be considered with utmost care for welding robot's safe operation. Twelve of these factors are shown in Figure 5.3 [7,17].

The factor "Explosion-proof switches" is concerned with using such switches whenever possible. The factor "area contamination" is concerned with ensuring that welding robots do not contaminate areas where spatter or harmful light rays may produce adverse effects. The factor "welding pollutants" is concerned with alleviating welding pollutants' problems. In this case, an electrostatic precipitator may help for reducing pollutants as well as the use of a centrally located charcoal collector for reducing welding gases. It is to be noted that this collector draws the residual gases through its torch attachment. Thus, it is quite useful in situations where it is

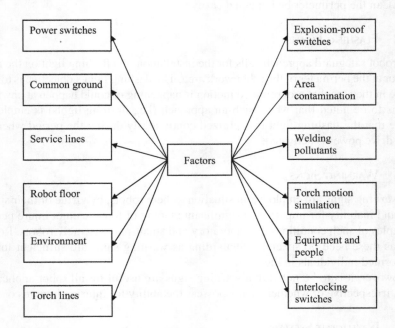

**FIGURE 5.3**   Factors to be considered for the safe operation of welder robots.

not possible to have an underfloor exhaust system as well as where the robot has to conduct many welds over fairly long distances.

The factor "torch motion simulation" is concerned with the welding torch motion's simulation prior to striking the arc. The factor "equipment and people" is concerned with preventing injury and damage to people and equipment, respectively. In this case, it is to be noted that by placing spatter shields and ultraviolet light-filtering screens around the welding table, the positioner and workstation can prevent injury and damage to people and equipment.

The factor "interlocking switches" is concerned with installing such switches to the welding jigs and fixtures, which highlight safe clamping. The factor "torch lines" is concerned with avoiding the entanglement of such lines with the joints of the robot during extreme movement. The items such as strings, counterweights, pivots, and pulleys can be used for eliminating this problem. The factor "environment" is concerned with discovering an area that is free from high humidity, vibration, dust, oil, and smoke and also has low traffic.

The factor "robot floor" is concerned with selecting a suitable floor and surrounding directly beneath the robot. In this case, it is to be noted that metal plates and explosion-proof concrete offer a very good solution for problems caused by sparks and hot metal falling to the floor. The factor "service lines" includes air, exhaust, water, ventilation, and electricity lines. Here, the aim should be to place such service lines underneath the floor or in floor channels away from the welding torch and torch lines, in addition to placing them away from the robot's work area. Furthermore, avoid having air and water lines in the same channel.

The factor "common ground" calls for not sharing the common ground in conditions where a welding robot is joining conventional welding systems with high-frequency power supplies. Finally, the factor "power switches" is concerned with locking such switches to the control panel and welding power supply.

As there are specific hazards and safeguards associated with robotized resistance welding, laser welding, and gas-shielded arc welding, each of these topics is described below, separately [4,7,17].

### 5.8.1 Hazards and Safety Measures for Robotized Resistance Welding

Resistance welding is the type of welding in which, at some point during the process, force is put to the surfaces in contact, and in which welding heat is generated by sending an electric current through the resistance at, and adjacent to, these surfaces [4,7,17]. The use of robots is quite useful for minimizing some of the hazards that are traditionally associated with resistance welding.

These associated hazards may be grouped under the following three classifications [7]:

*   **Classification I: Mechanical hazards**: Two examples of mechanical hazards are trapping by work-handling equipment and shocks from voltage conductors. The safety measures applicable to the robotized operation's normal mode are providing satisfactory guarding/interlocks for stopping

access to the hazard zone and balance support for minimizing mechanical stress on the robot arm.

- **Classification II**: **Electrical hazards**: These hazards are high/medium voltage, access to welding control equipment, and magnetic fields. The safety measures applicable to the robot operation's normal mode are the identification of workers wearing pacemakers, locking for stopping unauthorized access, and placing primary supply and welding transformers in a safe location.
- **Classification III**: **Other hazards**: These hazards are fire, fumes, and noise. Some of the safety measures to the robotized operation's normal mode are keeping closed containers of weld sealers and silencers on the air exhaust and placing extractor fans in an "on" position.

### 5.8.2   Robotized Laser Welding Hazards and Safety Measures

Laser welding is the type of welding where energy is produced by a laser beam and focused at a point. The hazards that emanate from the robotized laser welding's use are those connected with fumes and electromagnetic radiation. Welding lasers can cause fire as well as hazards to both eyes and skin (diffuse reflections).

When the robot is in its normal operation mode, there are many safety measures associated with power lasers. Seven of these safety measures are presented below [7,17].

- **Warning signs**: These are to be located at accesses to the enclosure for their effectiveness.
- **Emission indicator**: It highlights that the laser is in its active mode.
- **Remote interlock**: It is employed for stopping access to the power laser enclosure during laser action.
- **Key control**: It is useful for stopping unauthorized use.
- **Beam attenuator**: It helps to stop light from the laser entering the enclosure while individuals are still inside.
- **Training**: It is quite helpful for familiarization with the product and for hazard control.
- **Enclosure**: It helps to stop exposure to laser light.

### 5.8.3   Hazards and Safety Measures for Robotized
### Gas-Shielded Arc Welding

Gas-shielded arc welding is the type of welding where both the arc and molten pool are shielded from the atmosphere by gas [17]. When this type of welding is robotized, there are various types of associated hazards. Some of these hazards are resulting fumes, fire from hot metal welded parts, hot metal sparks, mechanical hazards, electric shock from the torch, and electromagnetic radiation [7,17]. For modes normal, programming, and maintenance solutions to these hazards are available in Ref. [17].

For the robotized operations' normal mode, selected hazards associated with gas-shielded arc welding and their corresponding safety measures (given in parentheses) are presented below [7]:

- Electric shock from the torch (place torch inside the work perimeter).
- Magnetic radiation (minimize this radiation to a normal level; computer enclosures should furnish an appropriate shield to electronics).
- Sparks from hot metal (position an interior perimeter guard and ensure that there is no flammable material inside the guarded workspace. Also, keep all cables and pipes inside the guarded workspace appropriately protected).
- Phosgene from chlorinated solvents on work items (dry all parts carefully after any cleaning process before welding).
- Fire from hot metal sparks (ensure that no flammable materials exist inside the perimeter guard and that all pipes and cables within the perimeter guard framework are appropriately protected).
- Ozone from the arc (provide proper fume extraction fans in the general area of work).
- Smoke from leftover oil on work items (minimize oil on the work surface to lower oil fumes as well as to ensure good welds).
- Fire from hot metal welded parts (ensure that all the unloaded items are properly kept away from all flammable materials).

## 5.9  PROBLEMS

1. What are the unique robot safety-related problems faced by safety professionals concerned with robots? List at least seven of such problems.
2. What are the basic types of robot hazards? Discuss each of these hazard types in detail.
3. Discuss robot manufacturers' and users' roles with respect to robot safety.
4. Discuss safety considerations in the robot design phase.
5. Discuss safety considerations in the robot operation and maintenance phase.
6. Discuss weak points in planning that can result in robot safety-related problems in an industrial setting.
7. List at least six weak points in design in regard to robot safety.
8. List at least 12 common robot safety-related features along with their corresponding intended functions.
9. Describe the following two robot safeguard approaches:
   - Intelligent systems.
   - Physical barriers.
10. What are the factors to be considered for the safe operation of welder robots?

## REFERENCES

1. Japanese Industrial Safety and Health Association, *An Interpretation of the Technical Guidance on Safety Standards in the Use, Etc., of Industrial Robots*, Tokyo, 1985.
2. American National Standards Institute, *American National Standard for Industrial Robots and Robot Systems-Safety Requirements*, ANSI/RIA R15.06-1986, New York, 1986.

3. Ziskovsky, J.P., Working Safely with Industrial Robots, *Plant Engineering*, May 1984, pp. 81–85.
4. Dhillon, B.S., *Robot System Reliability and Safety: A Modern Approach*, CRC Press, Boca Raton, Florida, 2015.
5. Ziskovsky, J.P., Risk Analysis and the Factor, Proceedings of the Robots 8th Conference, Vol. 2, June 1984, pp. 15.9–15.21.
6. Addison, J.H., *Robotic Safety Systems and Methods: Savannah River Site*, Report No. DPST-84-907 (DE 35-008261), December 1984, issued by E.I. du Pont de Nemours and Company, Savannah River Laboratory, Aiken, South Carolina 29808.
7. Dhillon, B.S., *Robot Reliability and Safety*, Springer-Verlag, New York, 1991.
8. Russell, J.W., Robot Safety Considerations: A Checklist, *Professional Safety*, December 1983, pp. 36–37.
9. Ramachandran, V., Vajpayee, S., Safety in Robotic Installations, *Robotics and Computer-Integrated Manufacturing*, Vol. 3, 1987, pp. 301–309.
10. Jiang, B.C., Robot Safety: Users' Guidelines, in *Trends in Ergonomics/Human Factors III*, edited by W. Karwowski, Elsevier, Amsterdam, 1986, pp. 1041–1049.
11. Bellino, J.P., Meagher, J., Design for Safeguarding, Proceedings of Robots East Seminar, Boston, Massachusetts, October 1985, pp. 24–37.
12. Nicolaisen, P., Ways of Improving Industrial Safety for the Programming of Industrial Robots, Proceedings of the 3rd International Conference on Human Factors in Manufacturing, November 1986, pp. 263–276.
13. Akeel, H.A., Intrinsic Robot Safety, Working Safely with Industrial Robots, edited by P.M. Strubhar, *Robotics International of the Society of Manufacturing Engineers*, Publications Development Department, One SME Drive, P.O.Box 930, Dearborn, Michigan, 1986, pp. 61–68.
14. Clark, D.R., Lehto, M.R., *Reliability, Maintenance, and Safety of Robots, Handbook of Industrial Robotics*, edited by S.Y. Nof, John Wiley and Sons, New York, 1999, pp. 717–753.
15. Bararett, R.J., Bell, R., Hudson, P.H., Planning for Robot Installation and Maintenance: A Safety Framework, Proceedings of the 4th British Robot Association Annual Conference, 1981, pp. 18–21.
16. Marton, T., Pulaski, J.L., Assessment and Development of HF Related Safety Designs for Industrial Robots and Robotic Systems, Proceedings of the Human Factors Society 31st Annual Meeting, 1987, pp. 176–180.
17. MTTA, *Safeguarding Industrial Robots, Part II: Welding and Allied Processes*, The Machine Tool Trades Association (MTTA), London, 1985.

# 6 Safety in Nuclear Power Plants

## 6.1 INTRODUCTION

Nuclear safety defined by the International Atomic Energy Agency (IAEA) as "The achievement of proper operating conditions, prevention of accidents or mitigation of accident consequences, resulting in protection of workers, the public and the environment from undue radiation hazards" [1–3]. This covers nuclear power plants as well as all other nuclear-related facilities and areas.

Over the years, due to the occurrence of various types of accidents in nuclear power plants/facilities, safety in nuclear power plants has become a very important issue, and it requires a continuing quest for excellence. All involved individuals and organizations need constantly to be alert to opportunities for reducing risks to the lowest level possible. Over the years, the nuclear power industrial sector around the globe has considerably improved the safety of reactors and other associated systems and has proposed safer reactor and related systems designs. However, it is not possible for guaranteeing perfect safety because of potential sources of problems such as human errors and external events that can have a much higher impact than initially anticipated.

This chapter presents various important aspects of safety in nuclear power plants.

## 6.2 SAFETY OBJECTIVES OF NUCLEAR POWER PLANTS

As per IAEA, three safety objectives of nuclear power plants are as follows [4]:

- **Objective I: General nuclear safety objective**: This objective is concerned with protecting society, individuals, and the environment by establishing and maintaining in all nuclear power plants a highly effective defence against radiological-associated hazards.
- **Objective II: Technical safety objective**: This objective is concerned with preventing with high confidence the accidents' occurrence in nuclear power plants; ensuring that all accidents' occurrence is appropriately taken into account during the design process of the plant, even those with rather very low occurrence probability and whose radiological consequences, if any, would be minor; and ensuring that the occurrence probability of severe accidents with serious radiological consequences is extremely low.
- **Objective III: Radiation protection objective**: This objective is concerned with ensuring in normal operation that radiation exposure within the plant boundary as well as due to any release of radioactive material from the

DOI: 10.1201/9781003212928-6

plant is as low as possible, economic and social-associated factors being taken into consideration appropriately, and ensuring proper mitigation of the extent of radiation-associated exposure due to accidents.

## 6.3   NUCLEAR POWER PLANT FUNDAMENTAL SAFETY PRINCIPLES

There are three types of nuclear power plant fundamental safety principles concerning technical issues, management, and defence in depth. Each of these types is described below, separately [4].

### 6.3.1   GENERAL TECHNICAL PRINCIPLES

There are many underlying technical principles concerning technical issues that are important for the successful application of safety-related technology in nuclear power plants. Seven of these principles are presented below [4].

- **Principle I: Radiation protection. In this case principle**: A system of radiation protection-associated practices, clearly consistent with recommendations of the IAEA and the International Commission on Radiological Protection (ICRP), is followed during the nuclear power plants' design, commissioning, operational, and decommissioning phases.
- **Principle II: Proven engineering practices. In this case principle**: Nuclear power technology is based on engineering-associated practices that are clearly proven by testing and experience, and which are clearly reflected in approved codes and standards as well as in other documented statements.
- **Principle III: Human Factors. In this case principle**: Individuals engaged in activities bearing on nuclear power plant safety are trained and qualified for conducting their duties. The probability/possibility of the human error occurrence in nuclear power plant operation is taken into consideration by facilitating right decisions by all involved operators and inhibiting wrong decisions and by providing proper mechanisms for detecting and correcting or compensating for error.
- **Principle IV: self-assessment: In this case principle**: Self-assessment for all important activities at a nuclear power plant ensures the involvement of individuals carrying out line functions in detecting problems that concern safety and performance and overcoming them.
- **Principle V: Safety assessment and verification: In this case principle**: Safety assessment is conducted prior to the start of a nuclear power plant's construction and operation. The assessment is well documented and reviewed independently. This assessment is subsequently updated in light of a significant amount of new safety-associated information.
- **Principle VI: Quality assurance. In this case principle**: Quality assurance is properly applied throughout activities at a nuclear power plant as an element of a comprehensive system for ensuring with high confidence that all items delivered, services, and tasks conducted satisfy stated requirements.

- **Principle VII: Peer reviews. In this case principle**: Independent peer reviews provide access for practices and programs used at nuclear power plants performing well and allow their usage at other nuclear power plants.

## 6.3.2 MANAGEMENT PRINCIPLES

There are three fundamental management principles that are, directly or indirectly, concerned with the establishment of a safety culture, the operating organization's responsibilities, and the provision of regulatory control and verification of safety-related activities. Each of these items is described below [4].

- **Principle I: Safety culture. In this case principle**: An established safety culture governs the interactions of all organizations and individuals involved in activities concerning nuclear power.
- **Principle II: Operating organization's responsibility. In this case principle**: The ultimate responsibility concerning a nuclear power plant's safety totally rests with the operating organization, and this is in no way diluted by the regulators', constructors', suppliers', contractors', and designers' separate responsibilities and activities.
- **Principle III: Regulatory control and independent verification. In this case principle**: The government establishes an appropriate legal framework for the nuclear industrial sector as well as an independent regulatory body for licensing and regulatory control of nuclear power plants and for enforcing the necessary regulations. Furthermore, it is absolutely essential that the separation between the responsibilities of the regulatory body and those of all other involved parties is absolutely clear, so that the involved regulators retain their total independence as a safety authority and are totally protected from undue pressure.

## 6.3.3 DEFENCE IN DEPTH PRINCIPLES

"Defence in depth" is singled out among the fundamental principles since it underlies the nuclear power plant's safety-related technology. Two corollary principles of "defence in depth" are accident prevention and accident mitigation. The general statement of "defence in depth", accident prevention, and accident mitigation are presented below [3,4].

- **Principle I: Defence in depth. In this case principle**: To compensate for potential human- and mechanical-associated failures, an appropriate defence in-depth concept is implemented, centred on various levels of protection including successive barriers for preventing the release of radioactive material to the surrounding environment.

  The concept incorporates protection of barriers by averting damage to the plant as well as to the barriers themselves. Furthermore, it also incorporates further measures to protect the people as well as the environment from harm in case the barriers in question are not totally effective.

- **Principle II: Accident prevention. In this case principle**: Principal emphasis is placed on the primary means of achieving safety effectively, which is the prevention of accidents' occurrence, particularly those which could lead to severe core damage.
- **Principle III: Accident mitigation. In this case principle**: Off-site and in-plant proper measures are available and are prepared, for that would considerably lower the effects of radioactive materials' accidental release.

## 6.4 NUCLEAR POWER PLANT SPECIFIC SAFETY-RELATED PRINCIPLES

Nuclear power plant safety objectives and fundamental safety principles provide a very good conceptual framework for the specific safety principles. The specific safety-related principles are concerned with the nuclear power plant emergency preparedness, accident management, siting, design, manufacturing, and construction, commissioning, operation, and decommissioning. The safety-related principles concerned with each of these eight areas are presented below, separately [4].

### 6.4.1 EMERGENCY PREPAREDNESS

Emergency planning and preparedness are comprised of actions needed for ensuring that, in the event of an accident's occurrence, all measures needed for protecting the public and the staff members could be conducted and that the use of such services would be disciplined appropriately.

Three safety principles that are concerned with emergency plans, emergency response facilities, and assessment of accident consequences and radiological monitoring are as follows [4]:

- **Principle I: Emergency plans. In this case principle**: Appropriate emergency plans are developed prior to the power plant's start-up and are exercised periodically for ensuring that all protection-associated actions can be implemented in the event of an accident occurrence which results in (or has the potential for) quite significant releases of radioactive-related material within as well as beyond the site boundary. All emergency planning zones defined around the power plant clearly permit the application of a graded response.
- **Principle II: Emergency response facilities. In this case principle**: For emergency response, a permanently equipped emergency centre off the site is available. Also, a similar centre on the site is provided for directing emergency-related actions within the power plant and for communicating with the off-site emergency agency/organization.
- **Principle III: Assessment of accident consequences and radiological monitoring. In this case principle**: All means are clearly available to all the responsible site staff members to be employed in early prediction of the

extent as well significance of any release of radioactive-associated materials if an accident were to take place, for quick and continuous assessment of the radiological condition, and for determining the need for protective actions.

## 6.4.2 ACCIDENT MANAGEMENT

Accident management as an element of accident prevention includes the measures to be conducted by operators during an accident sequence's evolution after conditions have come to exceed the plant's design but prior to the development of a severe accident. Similarly, accident management as an element of accident mitigation includes constructive measure by the involved operating staff members in the event of the occurrence of a severe accident, clearly directed to preventing the further progress of such an accident as well as alleviating its affects.

Three safety principles that are concerned with strategy for accident management, training and procedures for accident management, and engineered features for accident management are as follows [4]:

- **Principle I**: **Strategy for accident management. In this case principle**: The findings of an analysis of the response of the power plant to future accidents beyond the design basis are used properly in developing guidance on an accident management-associated strategy.
- **Principle II**: **Training and procedures for accident management. In this case principle**: All nuclear power plant staff members are properly trained as well as retrained in the procedures to be followed if an accident occurs that exceeds the plant's design basis.
- **Principle III**: **Engineered features for accident management. In this case principle**: All equipment, instrumentation, and diagnostic aids are available to all involved operators, who may, directly or indirectly, at some time be faced with the need for controlling the consequences and course of an accident that are well beyond the design basis.

## 6.4.3 SITING

The site is the area within which a nuclear power plant is located and is under the absolutely full control of the operating organization/company. The selection of a proper site for the nuclear power plant is a very important process as local circumstances can directly or indirectly affect safety considerably.

Four safety principles that are concerned with the external factors affecting the plant, radiological impact on the public and the local environment, feasibility of emergency plans, and ultimate heat sink provisions are as follows [4]:

- **Principle I**: **External factors affecting the plant. In this case principle**: The selection of the site clearly takes into consideration the investigations' findings of local factors that can adversely affect the power plant safety.

- **Principle II**: **Radiological impact on the public and the local environment. In this case principle**: The sites under consideration are properly investigated from the standpoint of the radiological impact of the power plant during normal operation as well as in accident conditions.
- **Principle III**: **Feasibility of emergency plans. In this case principle**: The site selected for a nuclear power plant is quite compatible with the off-site countermeasures that may be necessary for limiting the effects of the radioactive substances' accidental releases as well as is expected to remain compatible with such measures.
- **Principle IV**: **Ultimate heat sink provisions. In this case principle**: The site selected for a nuclear power plant contains a highly reliable long-term heat sink that can totally remove energy generated in the power plant after shutdown, both right after shutdown as well as over the longer period.

### 6.4.4  DESIGN

The primary objective of a nuclear power plant's designers is to provide a good design. These individuals ensure that the power plant's all structures, systems, and components have the appropriate characteristics, material composition, and specifications, and are integrated and laid out in such a way as to satisfy the general plant performance-associated specifications effectively.

The design-associated safety principles can be grouped under the following three areas [4]:

1. Design process.
2. General features.
3. Specific features.

The design process-associated safety principles are concerned with design management, proven technology, and general basis for design. Similarly, the general features-associated safety principles are concerned with plant process control systems, equipment qualification, automatic safety systems, radiation protection in design, inspection ability of safety equipment, reliability targets, and dependent failures.

Finally, the specific features-associated safety principles are concerned with the control of accidents within the design basis; start-up, shutdown, and low power generation; monitoring of plant safety status; reactor core integrity; confinement of radioactive material; plant physical protection, protection against power-transient accidents; new and spent fuel storage; automatic shutdown systems; preservation of control capability; station blackout; protection of confinement structure; reactor coolant system integrity; normal heat removal; and emergency heat removal.

### 6.4.5  MANUFACTURING AND CONSTRUCTION

In this case, a primary safety-related requirement is that a nuclear power station be manufactured and constructed as per the design intent. This is accomplished by maintaining proper attention to a variety of issues, from the broad aspect of account-ability of all involved organizations to the competence, diligence, and proper care of the involved individual personnel/workers.

Two safety principles that are concerned with the safety evaluation of design and the achievement of quality are presented below [4].

- **Principle I: Safety evaluation of design. In this case principle**: A nuclear power plant's construction is started only after the regulatory body and the operating organization/company have fully satisfied themselves through appropriate assessments that the major safety-associated issues have been fully resolved and that the remainder are amendable to solution prior to when operations are scheduled to take place.
- **Principle II: Achievement of quality. In this case principle**: The power plant manufacturers and constructors discharge their responsibilities for the provision of equipment and construction of good quality by using well-proven and developed methods and procedures clearly supported by quality assurance-associated practices.

### 6.4.6 COMMISSIONING

Commissioning is absolutely necessary for demonstrating that the completed power plant is satisfactory for service prior to it being made operational. For this very pur-pose, a well-planned and well-documented commissioning program is developed and executed. The operating organization/company, including its all potential oper-ating staff personnel, participate in this phase. Power plant systems are progressively handed over to the operating staff personnel as the installation and testing of each and every item are accomplished.

Four principles that are concerned with the verification of design and construc-tion, validation of operating and functional procedures, collecting baseline data, and preoperational plant adjustments are presented below [4].

- **Principle I: Verification of design and construction. In this case principle**: The commissioning program is developed and followed for demonstrating that the entire power plant, particularly items critical to safety and radiation protection, has been constructed and functions as per the design intent, and for ensuring that all weaknesses are detected and rectified.
- **Principle II: Validation of operating and functional test procedures. In this case principle**: All procedures involved for normal plant and systems operation and for functional tests to be conducted during the operational phase are appropriately validated as part of the commissioning program.
- **Principle III: Collecting baseline data. In this case principle**: During commissioning-associated tests, detailed diagnostic-associated data are collected on parts that have special safety-associated significance, and the system's initial operating parameters are documented.

- **Principle IV: Preoperational plant adjustments. In this case principle**:
  During the commissioning process, the as-built safety and process systems
  operating characteristics are appropriately determined and recorded. All
  operating points are adjusted for conforming to design-associated values
  as well as to safety analyses. All training-associated procedures, as well as
  limiting conditions for operation, are changed for reflecting accurately the
  systems as-built operating characteristics.

### 6.4.7 OPERATION

The operating organization/company is fully responsible for providing all equipment, staff, procedures, and management practices appropriate for safe operation,
including the fostering of an effective environment in which safety is clearly seen
as an important factor and a matter of personal accountability for all involved staff
members. It may seem on certain occasions that emphasis on safety might be in
conflict with the requirement for achieving a high-capacity factor and for satisfying
all demands of electricity generation. This very conflict is rather more apparent than
real as well as it can at most be transitory.

The operation-associated safety principles are concerned with safety review procedures; organization, responsibility, and staffing; radiation protection procedures;
quality assurance in operation; conduct of operations; normal operating procedures;
feedback of operating experience; training; engineering and technical support of
operations; operational limits and conditions; emergency operating procedures; and
maintenance, testing, and inspection.

Additional information on operation-associated safety principles is available in
Ref. [4].

### 6.4.8 DECOMMISSIONING

A nuclear power plant that is shut down still remains an operating plant until its decommissioning takes place and is still subject to normal control procedures and processes
for ensuring safety. After the termination of operations and the removal of spent fuel
from the plant, a significant radiation hazard still remains, which must be managed
appropriately for protecting the involved workers' health as well as the public.

The removal of power plant equipment and its decontamination can only be facil-
itated effectively if appropriate consideration is given during the design phase to
decommissioning as well as disposal of the wastes arising from the decommissioning process. The safety principle concerned with the decommissioning is presented
below [3,4].

- **Principle**: Appropriate consideration is given during design and plant
  operations for facilitating eventual decommissioning and waste management. After the termination of operations and the spent fuel's removal from
  the plant, radiation-associated hazards are managed so as to protect all
  involved workers' health as well as the public during the plant decommissioning process.

## 6.5    SAFETY MANAGEMENT IN NUCLEAR POWER PLANT DESIGN

The management of safety in nuclear power plant design is very important. In this regard, three requirements are presented below [3,5–7].

- **Requirement I**: **Management system for power plant design**. In this case, the design organization shall develop and implement a management which shall develop and implement a management system for ensuring that each and every safety-associated requirement developed for the design of the power plant is considered and implemented appropriately in all phases of the design process, and all these requirements are met effectively in the final design.
- **Requirement II**: **Safety of the power plant design throughout the lifetime of the power plant**. In this case, the power plant's operating organization/company shall develop a formal system for ensuring the continuing safety of the power plant design throughout the nuclear power plant's lifetime.
- **Requirement III**: **Responsibilities in the management of safety in power plant design**. In this case, an applicant for a licence for constructing and/or operating a nuclear power plant shall be fully responsible for ensuring that the design submitted to the regulatory body/organization clearly satisfies each and every applicable safety-related requirement.

## 6.6    DETERMINISTIC SAFETY ANALYSIS FOR NUCLEAR POWER PLANTS

Throughout the lifetime of a nuclear power plant, safety analyses play a very important role. The stages of and occasions in a nuclear power plant's lifetime in which the application of safety analyses is relevant are as follows [3,8]:

- Design.
- Commissioning.
- Operation and shutdown.
- Modification of design or operation.
- Periodic safety-associated reviews.
- Life extension in provinces/states where licences are issued for a limited period.

Two basic types of safety analysis are deterministic safety analysis and probabilistic safety analysis. Deterministic safety analyses for a nuclear power plant/facility predict the response to all postulated initiating events, and a specific set of rules and acceptance criteria is applied. Generally, they focus on neutronic, thermohydraulic, structural, and radiological aspects, which are often analyzed with different computational methods. The computations are generally conducted for predetermined operational states and operating modes, and the events incorporate expected postulated accidents, selected beyond design basis accidents, severe accidents with core degradation, and transients.

The options for conducting deterministic safety analyses are shown in Figure 6.1 [8].

Additional information on options are shown in Figure 6.1 is available in Ref. [8].

### 6.6.1 APPLICATION AREAS OF DETERMINISTIC SAFETY ANALYSIS

There are six areas in which deterministic safety analysis can be applied in nuclear power plants. These areas are as follows [3,8]:

- **Area I**: **Analysis of incidents that have taken place or of combinations of such incidents with all other hypothetical faults**: In this case, analyses would normally require best estimate approaches, in particular for sophisticated occurrences that need a realistic simulation.
- **Area II**: **Development and maintenance of emergency operating-associated procedures and accident management-associated procedures**: In this case, best estimate codes along with realistic assumptions should be employed.
- **Area III**: **Nuclear power plant's design**: In this case, deterministic safety analyses need either a conservative approach or a clearly best estimate analysis along with an evaluation of all uncertainties.
- **Area IV**: **Assessments by the regulatory body/organization of safety analysis-associated reports**: In this case, both conservative methods and best estimate plus uncertainty approaches may be employed.
- **Area V**: **Production of new or revised safety analysis documents for licensing, including obtaining the proper approval of the regulatory body/organization for changes to a power plant and to power plant operation**: In this case, both conservative methods and best estimate plus uncertainty approaches may be employed.

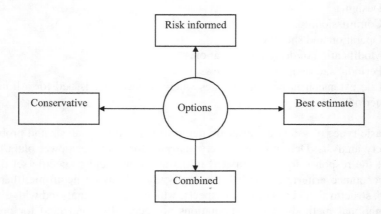

**FIGURE 6.1**   Options for conducting deterministic safety analyses in nuclear power plants.

- **Area VI**: **Refinement of earlier safety-associated analyses in the context of a periodic safety review for providing assurance that the earlier conclusions as well as assessments are still valid**.

## 6.7   SAFETY-ASSOCIATED REQUIREMENTS IN SPECIFIC NUCLEAR PLANT SYSTEMS' DESIGN

These requirements may be classified under the following ten areas [3,5]:

- **Area I: Instrumentation and control systems**: In this area, the requirements are concerned with the provision of instrumentation, control systems, protection systems, reliability and testability of instrumentation and control systems, use of computer-based equipment in systems important to safety, control room, separation of protection systems, and control systems, emergency control centre, and supplementary control room.
- **Area II: Reactor coolant systems**: In this area, the requirements are concerned with the design of reactor coolant systems, over pressure protection of the reactor coolant pressure boundary, heat transfer to an ultimate heat sink, removal of residual heat from the reactor core, inventory of reactor coolant, emergency cooling of the reactor core, and cleanup of reactor coolant.
- **Area III: Fuel handling and storage systems**: In this area, there is only one requirement, and it is concerned totally with fuel handling and storage systems.
- **Area IV: Supporting systems and auxiliary systems**: In this area, the requirements are concerned with performance of supporting systems and auxiliary systems, process sampling systems and post-accident sampling systems, fire protection systems, heat transport systems, overhead lifting equipment, air conditioning systems and ventilation systems, compressed air systems, and lighting systems.
- **Area V: Reactor core and associated features**: In this area, the requirements are concerned with the performance of fuel elements and assemblies, control of the reactor core, reactor shutdown, and structural capability of the reactor core.
- **Area VI: Other power conversion systems**: In this area, there is only one requirement, and it is concerned with the steam supply system, feedwater system, and turbine generators.
- **Area VII: Containment structure and containment system**: In this area, the requirements are concerned with the containment system for the reactor, isolation of the containment, access to the containment, control of containment conditions, and control of radioactive releases from the containment.
- **Area VIII: Radiation protection**: In this area, the requirements are concerned with design for radiation protection and means of radiation monitoring.

- **Area IX: Treatment of radioactive effluents and radioactive waste**: In this area, the requirements are concerned with systems for treatment as well as control of waste and systems for treatment and control of effluents.
- **Area X: Emergency power supply**: In this area, there is only one requirement, and it is totally concerned with emergency power supply.

The above ten areas' requirements are presented below [3,5].

- **Requirement 1**: This requirement is concerned with reliability and testability of instrumentation and control systems, and it is that at the nuclear power plant/facility, instrumentation and control systems for equipment/items important to safety shall be designed for very high reliability and periodic testability commensurate with the safety-associated function(s) to be carried out.
- **Requirement 2**: This requirement is concerned with the reactor coolant's cleanup, and it is that the proper facilities shall be allocated at the nuclear power plant for removing radioactive substances from the reactor coolant, including activated corrosion-associated products and fission-associated products derived from the fuel and all non-radioactive substances.
- **Requirement 3**: This requirement is concerned with fuel handling and storage systems, and it is that appropriate fuel handling and storage systems shall be provided at the nuclear plant/facility for ensuring that the fuel's integrity and properties are effectively maintained at all times during the fuel handling and storage process.
- **Requirement 4**: This requirement is concerned with heat transport systems, and it is that auxiliary systems shall be provided as required for removing heat from parts and systems at the nuclear power plant/facility that are highly essential to function in operational states as well as in accident conditions.
- **Requirement 5**: This requirement is concerned with reactor shutdown, and it is that the appropriate means shall be provided for ensuring that there is an effective capability for shutting down the nuclear plant's reactor in operational states as well as in accident conditions and that the shutdown condition can be kept even for the reactor core's most reactive conditions.
- **Requirement 6**: This requirement is concerned with lighting systems, and it is that adequate lighting shall be provided in all of a nuclear power plant's/facility's operational areas in operational states as well as in accident conditions.
- **Requirement 7**: This requirement is concerned with access to the containment and it is that access by operating staff members to the containment at a nuclear power facility shall be through airlocks appropriately equipped with doors that are interlocked for ensuring that at least one of the doors is closed during reactor power operation phase as well as in accident conditions.

- **Requirement 8**: This requirement is concerned with the means of radiation monitoring, and it is that the appropriate equipment shall be provided at the nuclear power plant/facility for ensuring that there is appropriate radiation monitoring in operational states and design basis accident conditions and, as far as is possible, in design extension conditions.
- **Requirement 9**: This requirement is concerned with systems for treatment and control of effluents, and it is that appropriate systems shall be provided at the nuclear power plant/facility for treating liquid and gaseous radioactive effluents for keeping their amounts well below the authorized limits on discharges and as low as reasonably possible to achieve.
- **Requirement 10**: This requirement is concerned with emergency power supply, and it is that at the nuclear power plant/facility, the emergency power supply shall be capable of supplying the needed power in expected operational occurrences and accident conditions in the event of the off-site power loss.

## 6.8   NUCLEAR POWER PLANT SAFETY-ASSOCIATED STANDARDS AND DOCUMENTS

Over the years, the IAEA has produced many standards/documents directly or indirectly concerned with the safety of nuclear power plants. Some of these standards/documents are presented below [3,9].

- Predisposal Management of Radioactive Waste from Nuclear Fuel Cycle Facilities, Series No. SSG-41, International Atomic Energy Agency, Vienna, Austria, 2016.
- Design of Electrical Power Systems for Nuclear Power Plants, Series No. SSG-34, International Atomic Energy Agency, Vienna, Austria, 2016.
- Safety Classification of Structures, Systems and Components in Nuclear Power Plants, Series No. SSG-30, International Atomic Energy Agency, Vienna, Austria, 2014.
- Commissioning for Nuclear Power Plants, Series No. SSG-28, International Atomic Energy Agency, Vienna, Austria, 2014.
- Safety of Nuclear Fuel Cycle Facilities, Series No. NS-R-5 (Rev. 1), International Atomic Energy Agency, Vienna, Austria, 2014.
- Periodic Safety Review for Nuclear Power Plants, Series No. SSG-25, International Atomic Energy Agency, Vienna, Austria, 2013.
- Establishing the Safety Infrastructure for a Nuclear Power Programme, Series No. SSG-16, International Atomic Energy Agency, Vienna, Austria, 2012.
- Disposal of Radioactive Waste, Series No. SSR-5, International Atomic Energy Agency, Vienna, Austria, 2011.
- Radiation Protection and Radioactive Waste Management in the Design and Operation of Research Rectors, Series No. NS-G-4.6, International Atomic Energy Agency, Vienna, Austria, 2009.

- Evaluation of Seismic Safety for Existing Nuclear Installations, Series No. NS-G-2.13, International Atomic Energy Agency, Vienna, Austria, 2009.
- Severe Accident Management Programmes for Nuclear Power Plants, Series No. NS-G-2.15, International Atomic Energy Agency, Vienna, Austria, 2009.
- The Management System for Nuclear Installations, Series No. GS-G-3.5, International Atomic Energy Agency, Vienna, Austria, 2009.
- Fundamental Safety Principles, Series No. SF-1, International Atomic Energy Agency, Vienna, Austria, 2006.
- Maintenance, Periodic Testing and Inspection of Research Reactors, Series No. NS-G-4.2, International Atomic Energy Agency, Vienna, Austria, 2006.
- Radiation Protection Aspects of Design for Nuclear Power Plants, Series No. NS-G-1.13, International Atomic Energy Agency, Vienna, Austria, 2005.
- Design of Reactor Containment Systems for Nuclear Power Plants, Series No. NS-G-1.10, International Atomic Energy Agency, Vienna, Austria, 2004.
- Seismic Design and Qualification for Nuclear Power Plants, Series No. NS-G-1.6, International Atomic Energy Agency, Vienna, Austria, 2003.
- Design of Fuel Handling and Storage Systems in Nuclear Power Plants, Series No. NS-G-1.4, International Atomic Energy Agency, Vienna, Austria, 2003.
- Organization and Staffing of the Regulatory Body for Nuclear Facilities, Series No. GS-G-1.1, International Atomic Energy Agency, Vienna, Austria, 2002.
- Maintenance, Surveillance and In-Service Inspection in Nuclear Power Plants, Series No. NS-G-2.6, International Atomic Energy Agency, Vienna, Austria, 2002.
- Modifications to Nuclear Power Plants, Series No. NS-G-2.3, International Atomic Energy Agency, Vienna, Austria, 2001.
- Fire Safety in the Operation of Nuclear Power Plants, Series No. NS-G-2.1, International Atomic Energy Agency, Vienna, Austria, 2001.
- Decommissioning of Nuclear Power Plants and Research Reactors, Series No. WS-G-2.1, International Atomic Energy Agency, Vienna, Austria, 1999.

## 6.9 PROBLEMS

1. Discuss safety objectives of nuclear power plants.
2. Discuss at least seven nuclear power plant fundamental safety principles related to technical issues.
3. Discuss the nuclear power plant safety principles with respect to the following two areas.

    (i)  Siting.

    (ii) Accident management.

4. Compare the nuclear power plant safety-related principles in the area of commissioning with those in the area of decommissioning.
5. Discuss the requirements with respect to safety management in nuclear power plant design.
6. What are the areas in which deterministic safety analysis can be applied in nuclear power plants?
7. List the areas in which safety-related requirements in specific nuclear plant systems' design may be classified.
8. Discuss the safety-associated requirements with the following two items:

    (i)  Fuel handling and storage systems.

    (ii) Reactor shutdown.

9. List at least 12 documents/standards directly or indirectly concerned with nuclear power plant safety.
10. Write an essay on safety in nuclear power plants.

## REFERENCES

1. *IAEA Safety Glossary: Version 2*, International Atomic Emergency Agency (IAEA), Vienna, Austria, 2007.
2. Nuclear Safety and Security, retrieved on September 7, 2016, from website: //en.wikipedia.org/wiki/Nuclear-safety-and-security.
3. Dhillon, B.S., *Safety, Reliability, Human Factors, and Human Error in Nuclear Power Plants*, CRC Press, Boca Raton, Florida, 2018.
4. *Basic Safety Principles of Nuclear Power Plants*, a report by the International Nuclear Energy Agency Advisory Group, Report No. 75-INSAG-3 Rev. 1: INSAG-12, International Atomic Energy Agency, Vienna, Austria, 1999.
5. *Safety of Nuclear Power Plants: Design*, Report No. SSR-2/1, International Atomic Energy Agency, Vienna, Austria, 2012.
6. *Maintaining the Design Integrity of Nuclear Installations Throughout Their Operating Life*, Report No. INSAG-19, International Atomic Energy Agency, Vienna, Austria, 2003.
7. *The Management System for Facilities and Activities*, IAEA Safety Standards Series No. GS-R-3, International Atomic Energy Agency, Vienna, Austria, 2006.
8. *Deterministic Safety Analysis for Nuclear Power Plants: Specific Safety Guide*, Report No. SSG-2, International Atomic Energy Agency, Vienna, Austria, 2009.
9. List of all Valid Safety Standards, retrieved on September 8, 2016, from website: http://www-ns.iaea.org/standards/documents/pubdoc-list.asp

# 7 Medical Systems Safety

## 7.1 INTRODUCTION

Each year a vast sum of money is spent around the globe to produce various types of medical systems/devices for use in the area of healthcare. A medical system/device must not only be reliable but also safe for users and patients. It is to be noted that the problem of safety concerning humans is not new; it can be traced back to the ancient Babylonian ruler Hammurabi. In 2000 BCE, Hammurabi developed a code known as the "Code of Hammurabi" with respect to health and safety [1–4]. The Code contained clauses in regard to injuries and financial damages against those causing injury to others.

In modern times, the passage of the Occupational Safety and Health Act (OSHA) by the U.S. Congress in 1970 is considered to be an important milestone with respect to health and safety in the United States. Two other important milestones that are specifically concerned with medical devices in the United States are the Medical Device Amendments of 1976 and the Safe Medical Device Act (SMDA) in 1990.

This chapter presents various important aspects of medical systems safety.

## 7.2 MEDICAL SYSTEMS SAFETY-RELATED FACTS AND FIGURES

Some of the facts and figures, directly or indirectly, concerned with medical systems safety are as follows [4–11]:

- After examining a sample of 15,000 hospital products, Emergency Care Research Institute (ECRI) concluded that around 4% to 6% of all these products were dangerous enough to warrant immediate corrective measure [5].
- In 1969, the special committee of the U.S. Department of Health, Evaluation, and Welfare reported that over a period of ten years, there were around 10,000 medical device-related injuries and 731 caused fatalities [6,7].
- As per Ref. [8], over time, ECRI has received many reports concerning radiologic equipment-related failures that either caused or had the potential to result in serious patient injury or death.
- A drug overdose took place due to incorrect advice given by an artificial intelligence medical system [9].
- A patient fatality took place due to a radiation overdose involving a Therac radiation therapy device [10].
- A five-year-old patient was crushed to death beneath the pedestal-style electric bed in which the child was placed after hospital admission [11].
- As per Ref. [8], faulty software programs in heart pacemakers caused two deaths.

DOI: 10.1201/9781003212928-7

## 7.3   MEDICAL DEVICE HARDWARE AND SOFTWARE SAFETY AND MEDICAL DEVICE SAFETY VERSUS RELIABILITY

Medical device/system hardware safety is very important because parts, such as electronic parts, are quite vulnerable to factors such as electrical interferences and environment-associated stresses. It simply means that each and every part in a medical device/system must be properly analyzed with respect to potential safety concerns and failures. For this very purpose, there are various methods/ approaches available to involved analysts including fault tree analysis (FTA) and failure modes and effect analysis (FMEA). Subsequent to part analysis, methods, such as part derating, safety margin, and load protection, can be employed for reducing the potential for the occurrence of failure of parts pinpointed as critical [3,12].

The safety of medical device software is as important as the safety of its hardware parts. However, it is to be noted that software in and of itself is not really unsafe, but the physical devices/systems it may control can directly or indirectly cause damage of varying degrees. For example, as per Ref. [12], an out-of-control software program can drive the gantry of a radiation therapy machine into a patient, or a hung software program may not only malfunction to stop a radiation exposure, but also deliver an overdose to a certain degree. All in all, it is to be noted with care that the software safety problem in medical devices is quite serious, as highlighted in a U.S. Food and Drug Administration (FDA) "device recalls" study [3,12]. More specifically, the study conducted over a period of six years (i.e., from 1983 to 1989) reported that 116 problems pertaining to software quality resulted in the recall of medical devices in the United States.

Although both safety and reliability are good things to which medical devices should aspire, from time to time there is some confusion, particularly in the industrial sector, with respect to the difference between medical device reliability and safety. Nonetheless, it is to be noted that reliability and safety are quite distinct concepts and at times they can have rather quite conflicting concerns [12].

A safe medical device/system may simply be expressed as a device/system that does not cause too much risk to property, humans, or equipment [3]. In turn, a risk is an undesirable event that can occur and is measured with respect to probability and severity. More simply, device/system safety is a concern with failures or malfunctions that introduce hazards and is expressed in terms of the level of risk, not in terms of satisfying specified requirements. On the other hand, a medical device/system reliability is the probability of success to satisfy its stated requirements.

Finally, it is to be noted that a medical device/system is still considered safe even if it often malfunctions without causing any mishap. In contrast, if a device/system functions normally at all times, but regularly puts directly or indirectly humans at risk, under this scenario the device/system is considered reliable but unsafe. Some examples of both these scenarios are available in Refs. [3,12].

## 7.4   TYPES OF MEDICAL DEVICE SAFETY AND ESSENTIAL SAFETY-ASSOCIATED REQUIREMENTS FOR MEDICAL DEVICES

Medical device safety may be classified under the following three categories or types [13]:

- **Category I: Descriptive Safety**: This type of safety is used in conditions when it is impossible or inappropriate for providing safety through the other two types/categories/means (i.e., conditional or unconditional). Nonetheless, descriptive safety with respect to operation, mounting, maintenance, transport, replacement, and connection may simply be statements such as "Not for Explosive Zones", "This Side Up", and "Handle with Care".
- **Category II: Conditional Safety**: This type of safety is used in conditions when unconditional safety cannot be realized. For example, in the case of an x-ray/laser surgical device, it is impossible for avoiding dangerous radiation emissions. However, it is well within means for minimizing risk with actions, such as incorporating a locking mechanism that permits device activation by authorized individuals only or limiting access to therapy rooms. Two examples of indirect safety-related means are protective laser glasses and X-ray folding screens.
- **Category III: Unconditional Safety**: This type of safety is preferred over all other possibilities or types because it is most effective. However, it calls for the eradication of all device-associated risks through design. However, it is to be noted with care that the use of warnings complements satisfactory device design, but does not replace it.

There are various types of, directly or indirectly, safety-associated requirements placed by the government and other agencies on medical devices. All these requirements may be grouped under the following three areas [4,13]:

- **Area I**: Safe design.
- **Area II**: Safe function.
- **Area III**: Sufficient information.

The requirements belonging to Area I (i.e., safe design) are mechanical hazard prevention, excessive heating prevention, care for environmental conditions, care for hygienic factors, protection against electrical shock, protection against radiation hazards, and proper material choice in regard to biological, mechanical, and chemical factors. Mechanical hazard prevention includes factors such as safe distances, breaking strength, and stability of the device. The excessive heating prevention-associated mechanisms are temperature control, effective design, and cooling. The care for environmental conditions includes factors such as humidity, temperature, and electromagnetic interactions. The remaining requirements are considered self-explanatory, however, the additional information on them is available in Ref. [13].

The components of Area II (i.e., safe function) are warning for or prevention of hazardous outputs, reliability, and accuracy of measurements. Finally, Area III (i.e., sufficient information) includes items such as effective labelling, instructions for use, packaging, and accompanying documentation.

## 7.5   SAFETY IN MEDICAL DEVICE LIFE CYCLE

Past experiences over the years clearly indicate that, in order to have safe medical devices, safety has to be considered throughout their life cycle. Thus, a medical device's life cycle may be divided into five phases, as shown in Figure 7.1 [3,14].

In the concept phase, past data and future technical-associated projections become the basis for the device under consideration and safety-related problems are highlighted and evaluated. The preliminary hazards analysis (PHA) method is quite an effective tool for identifying hazards during this phase. At the end of this phase, some of the typical questions to ask with respect to safety are as follows [3,14]:

- Are all the hazards highlighted and appropriately evaluated to develop hazard controls?
- Is the risk analysis initiated for developing mechanisms for hazard control?
- Are all the fundamental safety design-associated requirements for the phase in place so that the definition phase can be started?

The definition phase's main objective is providing proper verification of the initial design and engineering concerned with the medical device under consideration. The results of the PHA are updated along with the subsystem hazard analysis's initiation and their ultimate integration into the overall device hazard analysis. Methods, such as fault hazard analysis and FTA, may be employed for examining certain known hazards and their effects. All in all, the system definition will initially lead to the acceptability of a desirable general device design even though, because the design's incompleteness, not all related hazards will be completely known.

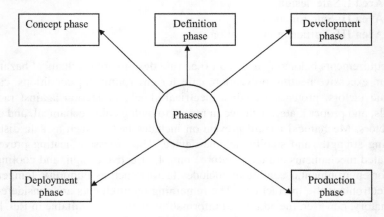

**FIGURE 7.1**   Medical device life cycle phases.

During the device's development phase, the efforts are directed on areas such as operational use, environmental impact, producibility engineering, and integrated logistics support. With the aid of prototype analysis and testing results, the comprehensive PHA is carried out for examining man–machine-related hazards, in addition to developing PHA further because of more completeness of the design of the device under consideration.

In the production phase, the device safety engineering report is prepared by utilizing the data collected during the phase. The report documents and highlights the device-related hazards. Finally, during the deployment phase, data concerning failures, accidents, incidents, etc., are collected, and safety professionals review any changes to the device. The device safety-related analysis is updated as necessary.

## 7.6 SOFTWARE-RELATED ISSUES IN CARDIAC RHYTHM MANAGEMENT PRODUCTS SAFETY

In regard to cardiac rhythm management products, there are many important issues to consider with care during their software safety-related analysis. These issues include marketing issues, technical issues, and management issues. It means that the developers of software must examine such issues with care in light of product context, constraints, and environmental requirements [3,4,15]. Two examples of cardiac rhythm management systems/products are defibrillators and pacemakers utilized for providing electrical therapy to malfunctioning cardiac muscles. Nonetheless, the marketing issues' four main elements are shown in Figure 7.2 [3,4,15]. These main

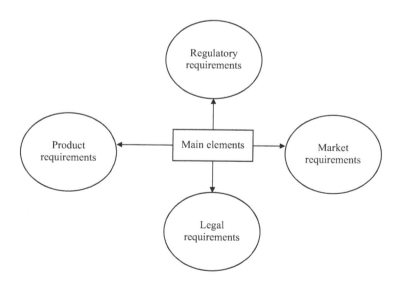

FIGURE 7.2   Main elements of the marketing issues.

elements are product requirements, regulatory requirements, market requirements, and legal requirements.

With respect to the product requirements, a typical cardiac rhythm management system/device is made up of electrical subsystems, advanced software, and mechanical subsystems that must function normally for its successful operation. Furthermore, as modern medical devices' software subsystem is composed of about 500,000 lines of code, its safety, efficiency, and reliability for controlling both internal and external operations are very important.

The regulatory requirements are important as regulatory bodies/agencies, such as the FDA, require that companies producing cardiac rhythm management systems must have systematic and rigorous software development processes, including safety analysis. It is estimated that around half of the device/product development cycle is directly or indirectly consumed by the regulatory acceptance process.

In regard to the market requirements, the cardiac rhythm management systems'/products' sheer size market is the sole important factor in their safety. For example, each year in the United States alone around half a million individuals experience a sudden cardiac death episode and about 31,000 individuals receive a defibrillator implant [3,4,15,16]. Furthermore, as per Ref. [15], the predictions for 1997 for the world market of cardiac rhythm management products/systems were approximately US$3 billion.

Finally, legal requirements also are very important because of the cardiac rhythm management systems'/devices' life-or-death nature, and the concerned regulations lead to highly sensitive legal requirements that involve regulatory bodies, manufacturers, patients and their families, etc. More clearly, in the event of injury to a patient or death, lawsuits by the regulatory bodies as well as the termination of devices/products by these bodies because of their safety-associated problems are the driving force for the legal requirements.

The technical issues are an important factor as well during the software development process because the software maintenance and complexity-associated concerns basically determine the analysis method and the incorporation process to be utilized. In regard to maintenance, the code resulting from the required modifications of safety-critical software is one of the most critical elements. A study conducted by one medical device software developer reported that out of the software-associated change requests, and subsequent to the internal release of software during development, 41% were concerned with software safety [3,15]. With respect to complexity, modern medical devices can have a quite large number of parallel, asynchronous, and real-time software tasks reacting to randomly occurring external events. Thus, for ensuring the timely and correct behaviour of such complex software is quite difficult, as is the mitigation and identification of safety faults with correctness and timeliness.

Finally, the management issues are primarily concerned with making appropriate changes to the ongoing development process for including explicit safety analysis, thus requiring convincing justifications and a clear vision. In this case, it will certainly be quite helpful if the management is shown explicitly that the cost of conducting software safety analysis during the development process can help to reduce the overall development cost, market losses, regulatory-associated problems, and avoid potential legal costs.

## 7.7   MEDICAL DEVICE ACCIDENT CAUSES CLASSIFICATIONS AND LEGAL ASPECTS OF MEDICAL DEVICE SAFETY

There are many causes for the occurrence of medical device-associated accidents. The professionals working in the area have classified these accident causes under the following seven classifications [17]:

- **Classification I**: Design defect.
- **Classification II**: Manufacturing defect.
- **Classification III**: Random component failure.
- **Classification IV**: Operator/patient error.
- **Classification V**: Faulty calibration, preventive maintenance, or repair.
- **Classification VI**: Abnormal or idiosyncratic patient response.
- **Classification VII**: Malicious intent or sabotage.

Additional information on the above seven classifications is available in Ref. [17].

One of the main objectives of system/product/device safety is to limit legal liability as much as possible. Tort law complements safety-related regulations through the deterrent of manufacturing harmful medical devices, in addition to providing satisfactory compensation to all injured persons. The decision of the U.S. Supreme Court on the Medtronic, Inc., *v.* Lohr case, for Lohr, put additional pressure on companies for producing reliable and safe medical devices [18]. The case was filed by a Florida woman, Lora Lohr, who had a cardiac pacemaker implanted, which was produced by Medtronic, Inc., for regulating her abnormal heart rhythm. The pacemaker malfunctioned, and she alleged that the device's malfunction was due to defective design, manufacturing, and labelling [18]. The three commonly used theories for making manufacturers liable for injury caused by their products are shown in Figure 7.3 [3,18]. The theories shown are a breach of warranty, negligence, and strict liability.

With respect to a breach of warranty, it may be alleged under the following three scenarios [3,18]:

- **Scenario I**: Breach of an expressed warranty.
- **Scenario II**: Breach of the implied warranty of suitability for a specific use.
- **Scenario III**: Breach of the implied warranty of merchantability.

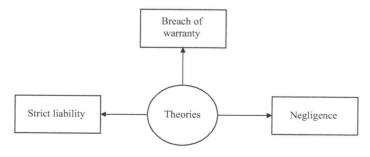

**FIGURE 7.3**   Commonly used theories for making manufacturers liable for injury caused by their products.

For example, if a medical device caused injury to an individual because of its inability to function as warranted, that device's manufacturer faces liability under the breach of an expressed warranty scenario (i.e., Scenario I).

In the case of negligence, if the device manufacturer fails to exercise reasonable care or fails for meeting a reasonable standard of care during the device manufacturing, handling, or distribution process, it could be liable for any damages resulting from the device. Finally, with respect to strict liability, the basis for imposing it is that the manufacturer of a device is in the best position for reducing related risks.

## 7.8   MEDICAL SYSTEM SAFETY ANALYSIS METHODS

There are many methods that can be used for performing safety analysis of medical systems/devices. Some of these methods are as follows [2,3,14,19–21]:

- Interface safety analysis.
- Operating hazard analysis.
- Preliminary hazard analysis.
- Human error analysis.
- Root cause analysis.
- Fault tree analysis.
- Technic of operations review.
- Failure modes and effect analysis.

Past experiences over the years clearly indicate that conducting an effective safety analysis of medical systems/devices requires a careful consideration in the selection and implementation of appropriate safety analysis methods for given situations. Thus, questions, such as those presented below, should be asked with care prior to selection and implementation of safety analysis methods for situations under consideration [3,19].

- What type of information, data, etc., is needed prior to the start of the study?
- What is the exact time frame for the initiation of analysis as well as its completion, submission, review, and update?
- When are the end results needed?
- Who are the end results users?
- What mechanism is required for acquiring information from subcontractors (if applicable)?

Three of the above eight methods are presented below and the remaining five methods are described in Chapter 4.

### 7.8.1   Operating Hazard Analysis

This method mainly focuses on hazards occurring from tasks/activities for operating system functions that take place as the system is used, transported, or stored. Generally, the operating hazard analysis (OHA) is initiated early in the system development cycle so that appropriate inputs to technical orders are provided, which in turn govern the system's testing. The application of the OHA provides a basis for safety considerations, such as follows [3,14,19]:

- Special safety procedures with respect to servicing, training, transporting, handling, and storing.
- Design modifications for eradicating hazards.
- Development of emergency procedures, warning, or special instructions with respect to operation.
- Safety guards and safety devices.
- Identification of system/item functions relating to hazardous occurrences.

It is to be noted that the analyst involved in the performance of OHA requires engineering descriptions of the system/device under consideration with appropriate available support facilities. Furthermore, OHA is conducted utilizing a form that needs information on items such as the operational event description, hazard description, hazard effects, hazard control, and requirements.

Additional information on this method is available in Refs. [14,19].

### 7.8.2  FAULT TREE ANALYSIS

This is a widely used method for performing safety and reliability analysis of engineering systems in the industrial sector. The method was originally developed in the early 1960s for evaluating the safety of the Minuteman Launch Control System [22]. Some of the main points concerned with fault tree analysis are as follows [3,4]:

- It is a very useful analysis in the early design phases of new systems/devices/items.
- It permits users for evaluating alternatives as well as pass judgement on acceptable trade-offs among them.
- It is an effective tool for analyzing operational systems/devices for desirable or undesirable occurrences.
- It can be used for evaluating certain operational functions (e.g., start-up or shutdown phases of system/device/facility operation).

Additional information on this method is available in Chapter 4 and in Ref. [22].

### 7.8.3  HUMAN ERROR ANALYSIS

This method is considered very useful for identifying hazards prior to their occurrence in the form of accidents. There could be the following two approaches to human error analysis:

- Conducting tasks for obtaining first-hand information on hazards.
- Observing all involved workers during their work hours in regard to hazards.

All in all, regardless of the performance of the human error analysis, it is strongly recommended to conduct it in conjunction with failure modes and effect analysis and hazard and operability (HAZOP) analysis methods presented in Chapter 4.

Additional information on this method is available in Refs. [2,14,19–21].

## 7.9 PROBLEMS

1. List at least five facts and figures concerned with medical systems safety.
2. Discuss medical device safety versus reliability.
3. Discus essential safety-associated requirements for medical devices placed by government and other agencies.
4. Discuss medical device hardware and software safety.
5. Discuss safety in a medical device life cycle.
6. What are the important software issues in cardiac rhythm management products safety?
7. What are the classifications of medical device accident causes?
8. What are the commonly used theories for making manufacturers liable for injury caused by their products?
9. Describe operating hazard analysis.
10. Write an essay on medical systems safety.

## REFERENCES

1. Ladou, J. Ed., *Introduction to Occupational Health and Safety*, The National Safety Council, Chicago, 1986.
2. Goetsch, D.L., *Occupational Safety and Health*, Prentice Hall, Englewood Cliffs, NJ, 1996.
3. Dhillon, B.S., *Medical Device Reliability and Associated Areas*, CRC Press, Boca Raton, Florida, 2000.
4. Dhillon, B.S., *Safety and Human Error in Engineering Systems*, CRC Press, Boca Raton, Florida, 2013.
5. Dhillon, B.S., Reliability Technology in Health Care Systems, Proceedings of the IASTED International Symposium on Computes, Advanced Technology in Medicine, and Health Care Bioengineering, 1990, pp. 84–87.
6. *Medical Devices, Hearing before the Subcommittee on Public Health and Environment*, U.S. Congress Interstate and Foreign Commerce, Serial No. 93-61, U.S. Government Printing Office, Washington, D.C., 1973.
7. Banta, H.D., The Regulation of Medical Devices, *Preventive Medicine*, Vol. 19, 1990, pp. 693–699.
8. *Mechanical Malfunctions and Inadequate Maintenance of Radiological Devices*, Medical Device Safety Report, Prepared by the Emergency Care Research Institute, Plymouth Meeting, PA, 2001.
9. Schneider, P., Hines, M.L.A., Classification of Medical Software, Proceedings of the IEEE Symposium on Applied Computing, 1990, pp. 20–27.
10. Casey, S., *Set Phasers on Stun: And Other True Tales of Design Technology and Human Error*, Aegean, Inc., Santa Barbara, CA, 1993.
11. *Electric Beds Can Kill Children, Medical Device Safety Report*, prepared by the Emergency Care Research Institute, Plymouth Meeting, PA, 2001.

12. Fries, R.C., *Reliable Design of Medical Devices*, Marcel Dekker, New York, 1997.
13. Leitgeb, N., *Safety in Electromedical Technology*, Interpharm Press, Buffalo Grove, IL, 1996.
14. Roland, H.E., Moriarty, B., *System Safety Engineering and Management*, John Wiley and Sons, New York, 1983.
15. Mojdehbakhsh, R., Tsai, W.T., Kirani, S., Elliott, L., Retrofitting Software Safety in an Implantable Medical Device, *IEEE Software*, Vol. 11, January 1994, pp. 41–50.
16. Lowen, B., Cardiovascular Collapse and Sudden Cardiac Death, in *Heart Disease: A Textbook of Cardiovascular Medicine*, edited by E. Braunwald, W.B. Saunders, Philadelphia, 1984, pp. 778–803.
17. Brueley, M.E., Ergonomics and Error: Who is Responsible? Proceedings of the First Symposium on Human Factors in Medical Devices, 1989, pp. 6–10.
18. Bethune, J., Ed., On Product Liability: Stupidity and Waste Abounding, *Medical Device and Diagnostic Industry Magazine*, Vol. 18, No. 8, 1996, pp. 8–11.
19. System Safety Analytical Techniques, *Safety Engineering Bulletin*, No. 3, May 1971. Available from the Electronic Industries Association, Washington, D.C.
20. Gloss, D.S., Wardle, M.G., *Introduction to Safety Engineering*, John Wiley and Sons, New York, 1984.
21. Hammer, W., *Product Safety Management and Engineering*, Prentice Hall, Englewood, NJ, 1980.
22. Dhillon, B.S., Singh, C., *Engineering Reliability: New Techniques and Applications*, John Wiley and Sons, New York, 1981.

# 8 Airline and Ship Safety

## 8.1 INTRODUCTION

Nowadays, airlines and ships are important modes of transportation around the globe. The world's 900 airlines with a total of approximately 22,000 aircraft, each year carry over 1.6 billion passengers for leisure and business travel, and around 40% of world trade of goods is carried by air [1–4]. Similarly, there are around 90,000 merchant ships in the world, and they transport over 90% of the world's cargo [4–6]. Over the years, airline and ship safety has been a very important issue, and for its improvement, various measures have been taken. For example, in the United States in the area of civil aviation, the Air Commerce Act was passed in 1926 [7,8].

The act required the examination and licensing of pilots and aircraft, appropriate investigation of accidents, as well as the establishment of safety rules. Due to measures such as these, the safety in the airline area has improved quite significantly, and currently the accident rate for air travel is approximately one fatality per 1 million flights [1–3].

In the sea transportation area, over the years there have been many accidents and other safety-associated problems. For example, the sinking of the RMS Titanic, a passenger liner operated by a British shipping company, in 1912 resulted in 1,517 onboard deaths [9]. Over the years, safety in the area of sea transportation has improved quite significantly, but it is still a very important issue.

This chapter presents various important aspects of airline and ship safety.

## 8.2 UNITED STATES AIRLINE-ASSOCIATED FATALITIES AND ACCIDENT RATE

The history of airline-associated crashes in the United States goes back to 1926 and 1927, when there were a total of 24 fatal commercial airline crashes. In 1929, there were 51 airline crashes that caused 61 deaths, and this remains the worst year on record, with an accident rate of around one per 1 million miles flown [8,10]. Over the years, airline safety in the United States has improved quite dramatically, but many airline-associated deaths still occur.

For the period 1983–1995, the number of deaths due to commercial airline accidents in the United States was as follows [4,8]:

- 1983: deaths: 8.
- 1984: deaths: 0.
- 1985: deaths: 486.
- 1986: deaths: 0.
- 1987: deaths: 212.

DOI: 10.1201/9781003212928-8

- 1988: deaths: 255.
- 1989: deaths: 259.
- 1990: deaths: 8.
- 1991: deaths: 40.
- 1992: deaths: 25.
- 1993: deaths: 0.
- 1994: deaths: 228.
- 1995: deaths: 152.

Accident rates per million flight departures for 1995, 1994, 1993, 1992, 1991, 1990, and 1989 were 0.40, 0.27, 0.28, 0.22, 0.33, 0.29, and 0.37, respectively [8].

It is to be noted that in comparison to fatalities in other sectors, the airline-related fatalities are extremely low. For example, in 1995 people were around 300 times more likely to die in a motor-vehicle-associated accident and around 30 times more likely to get drowned than to get killed in an airplane-related accident [8].

## 8.3   AIRCRAFT ACCIDENTS DURING FLIGHT PHASES AND CAUSES OF AIRPLANE CRASHES

A flight phase may be divided into the following nine distinct subphases [11]:

- Subphase I: ramp/taxi.
- Subphase II: take off.
- Subphase III: initial climb.
- Subphase IV: climb.
- Subphase V: cruise.
- Subphase VI: descent.
- Subphase VII: initial approach.
- Subphase VIII: final approach.
- Subphase IX: landing.

Past experiences over the years indicate that the accidents' occurrence can vary quite considerably from one flight subphase to another. For example, during the period 1987–1996, the highest percentage of aircraft accidents occurred during the final approach subphase and the lowest during the ramp/taxi subphase [11,12].

More specifically, the rough breakdowns of the percentages of accidents during 1987–1996 for the subphases were as follows [11,12]:

- 1% (ramp/taxi).
- 5% (cruise).
- 7% (descent).
- 8% (climb).
- 10% (initial climb).
- 11% (initial approach).

- 14% (take off).
- 21% (landing).
- 23% (final approach).

Finally, it is to be noted that accident data for the period 1990–1999 exhibit a similar trend [11,13].

Past experiences over the years indicate that there are many causes of airplane crashes. For example, a study of 19 major crashes (defined as one in which at least 10% of the airplane passengers are killed) of U.S. domestic jets occurring during the period 1975–1994 has highlighted eight different causes of all these crashes [4,8,14]. These causes (with a corresponding number of crashes in parentheses) were as follows [8,14]:

- Thunderstorm wind shear (4).
- Ground or air collisions (3).
- Ice buildup (3).
- Engine loss (2).
- Taking off without the flaps in the right position (2).
- Hydraulic failure (2).
- Cause unknown (2).
- Sabotage (1).

## 8.4 AIR SAFETY REGULATORY BODIES AND THEIR RESPONSIBILITIES

In the United States, there are two bodies (i.e., the Federal Aviation Administration (FAA) and the National Transportation Safety Board (NTSB)) that serve as the public's watchdog for safety in the aviation industrial sector. The history of these two agencies/bodies goes back to 1940 when the Civil Aeronautics Authority was split into two organizations: Civil Aeronautics Board (CAB) and Civil Aeronautics Administration (CAA).

Since then, CAB has evolved into the National Transportation Safety Board (NTSB) and, similarly, CAA has evolved into the Federal Aviation Administration (FAA). Now, both NTSB and FAA are part of the U.S. Department of Transportation [8]. The current responsibilities of both NTSB and FAA are presented in the following subsections.

### 8.4.1 NATIONAL TRANSPORTATION SAFETY BOARD RESPONSIBILITIES

The NTSB has responsibilities inside and outside the aviation industrial sector. More clearly, in addition to investigating aviation-associated accidents, the NTSB is also responsible for investigating significant accidents occurring in other modes of transportation, such as railroad and marine. Nonetheless, the four main responsibilities of the NTS are as follows [8]:

- Conduct special studies on transportation safety-associated issues.
- Serve as the "court of appeals" for FAA-related matters.
- Issue appropriate safety recommendations to help prevent the potential accidents' occurrence.
- Maintain the government database on aviation-associated accidents.

### 8.4.2 FEDERAL AVIATION ADMINISTRATION RESPONSIBILITIES

The FAA has many responsibilities. Its main responsibilities are as follows [8,15]:

- Establishing airline safety-associated regulations.
- Developing operational requirements for airlines.
- Reviewing the design, manufacture, and maintenance of aircraft-related equipment.
- Developing, operating, and maintaining the nation's air control system.
- Establishing minimum standards for crew training.
- Conducting safety-related research and development.

## 8.5 AVIATION RECORDING AND REPORTING SYSTEMS

There are various types of aviation recording and reporting systems used around the globe. In the United States, there are four major organizations that collect and analyze aviation safety-associated data. These organizations are the Federal Aviation Administration, the National Aeronautics and Space Agency (NASA), the Research and Special Programs Administration (RSPA), and the National Transportation Safety Board (NTSB). Nonetheless, some of the data systems that can be quite useful for obtaining aviation safety-related information are as follows [4,7]:

- Aviation Safety Reporting System.
- Aviation Accident/Incident Reporting System.
- Aviation Safety Analysis System.
- International Civil Aviation Organization's (ICAO) Accident/Incident Reporting System (ADREP).
- Pilot Deviation Data base.
- Air Operator Data System.
- Air Carrier Statistics Database.
- NTSB Accident/Incident Reporting System.
- Operational Error Database.
- Accident Incident Data System.
- Air Transportation Oversight System (ATOS).
- Service Difficulty Reporting System.
- Near-Midair Collision Database.

Six of these data systems are described separately in the following subsections.

### 8.5.1   ACCIDENT/INCIDENT REPORTING SYSTEM

This system was developed by the NTSB, and it contains information on all known civil aviation-related accidents that have occurred in the United States over the years. This database/system may simply be referred to as the official repository of aviation-related accident data and causal factors.

The database data are classified under the following nine categories [7]:

- Category I: Location information.
- Category II: Aircraft information.
- Category III: Sequence of events.
- Category IV: Operator information.
- Category V: Narrative.
- Category VI: Findings.
- Category VII: Weather/environmental information.
- Category VIII: Injury summary.
- Category IX: Pilot information.

Additional information on this database/system is available in Ref. [7].

### 8.5.2   ACCIDENT INCIDENT DATA SYSTEM (AIDS)

This system was developed by the FAA, and it contains incident data records for all categories of civil aviation in the United States. More specifically, AIDS contains air-carrier-associated and general-aviation-associated incidents from 1978 as well as general-aviation-related accidents from 1973 [7]. Incidents in this database are events that do not meet the aircraft damage or human-injury thresholds contained in the NTSB's definition of an accident.

The information contained in the data system is obtained from various sources, including incidents reports on FAA Form 8020-5[7]. The data in the system are classified under seven categories shown in Figure 8.1 [7].

Additional information on this database/data system is available in Ref. [7].

### 8.5.3   AVIATION SAFETY ANALYSIS SYSTEM

This system was developed by FAA, and its databases fall under the four categories shown in Figure 8.2 [4,7]. These categories are operational data, regulatory data, organizational information, and airworthiness data.

Operational data concerned with the aviation-related environment, such as tracking aircrew, operators, and aircraft along with incidents, accidents, enforcement actions, and mechanical reliability reports. Regulatory data are concerned with background information, such as notices of proposed rulemaking, previous regulations, and legal opinions. Organizational information is concerned with the work management subsystems for monitoring aviation standards tasks, such as airline

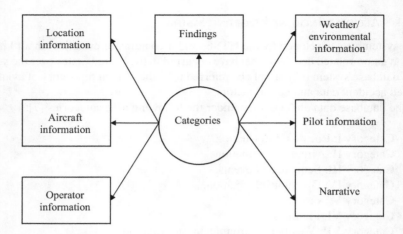

**FIGURE 8.1**    Accident incident data system database categories.

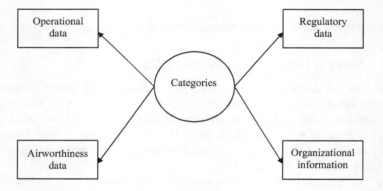

**FIGURE 8.2**    Aviation safety analysis system database categories.

inspections. Finally, airworthiness data are mainly concerned with historical-related information on aircraft, such as FAA-specified mandatory modifications.

### 8.5.4  AVIATION SAFETY REPORTING SYSTEM

This system is the result of a joint effort of FAA, Battelle Memorial Institute, and National Aeronautics and Space Administration (NASA), and it is maintained at Battelle Laboratories in Columbus, Ohio. This is a voluntary reporting system, where pilots, air traffic controllers, and others can submit accounts of aviation-associated incidents. The system became operational on April 15, 1976, and contains approximately 500,000 aviation-incident-associated reports to date.

The system's reporting form is specifically designed for gathering the maximum amount of information without discouraging reporters. Additional information on this system is available in Ref. [7].

## 8.5.5   ICAO ADREP System

The International Civil Aviation Organization's (ICAO's) Accident/incident Reporting System (i.e., ADREP) is a data bank that was established in 1970. It contains worldwide civil aviation accident- and incident-associated information of aircraft (fixed-wing and helicopter) heavier than 5,700 kg. Based on this database, ICAO provides the following information [7]:

- **ADREP Requests**. These are computer printouts that ICAO provides in response to specific requests from countries.
- **ADREP Annual Statistics**. This is an ICAO circular that contains annual statistics from the database.
- **ADREP Summary**. This summary contains the ADREP preliminary reports as well as data reports received by ICAO during a two-month period and is issued six times a year.

Additional information on this database is available in Refs. [7,16].

## 8.5.6   Air Transportation Oversight System (ATOS)

This system was implemented in 1998 as a modern approach to FAA certification and surveillance oversight, using system safety-associated principles and systematic processes for ensuring that the nation's all airlines are complying with FAA rules and regulations and have appropriate levels of safety built into their all operating systems. The system incorporates items such as the structured application of new inspection-associated tasks, analytical processes, and data-collection approaches into the oversight of individual airlines.

Under ATOS, operations of an airline are divided into 7 systems, 14 subsystems, and 88 underlying component elements, which provide a quite good structure for highlighting areas of concern or risks, collecting data, and conducting surveillance [7]. Additional information on this system is available in Ref. [7].

## 8.6   WORLDWIDE AIRLINE ACCIDENT ANALYSIS

Airlines are a widely utilized mode of transportation throughout the world. Currently, over 16,000 jet aircraft are being used around the globe, with over 17 million departures per year [7]. A study of worldwide scheduled commercial jet operations during the period 1959–2001 clearly indicates that there were a total of 1,307 accidents, resulting in 24,700 onboard deaths [7,17]. By type of operation, these 1,307 accidents can be grouped under the following three categories [7]:

- **Category I: Passenger operations**: 1,033 accidents (79%).
- **Category II: Cargo operations**: 169 accidents (13%).
- **Category III: Testing, training, demonstration, or ferrying**: 105 accidents (8%).

The collective U.S. and Canadian element of these 1,307 accidents was approximately 34% (i.e., 445 accidents), which contributed to around 25% (6,077) of the worldwide 24,700 onboard deaths [7]. A study of the 1959–2001 accident data indicates that the world commercial jet fleet accident rate (i.e., accidents per million departures) for the period 1974–2001 has been fairly stable [17].

Additional information on the subject is available in Refs. [7,17].

## 8.7   NOTEWORTHY MARINE ACCIDENTS

Over the years, there have been many marine accidents. Four of the more noteworthy of these accidents are described in the following four subsections.

### 8.7.1   PRESTIGE ACCIDENT

This accident is concerned with a 26-old Bahamian-registered and American Bureau of Shipping (ABS)-classed single-hull oil tanker, called *Prestige*. *Prestige* left Riga, Latvia, on November 5, 2002, with a cargo of 77,000 tons of heavy oil. On November 13, 2002, the tanker developed a substantial starboard list in heavy seas and in high winds in the region around 30 nautical miles off the coast of Galicia, Spain [18,19]. A large crack was found in the starboard side of the hull, and the vessel lost its main propulsion and started to drift.

The *Prestige*'s all 27 crew members were evacuated safely. On November 19, 2002, the oil tanker broke into two and sank around 133 nautical miles off the coast of Spain. The incident quite seriously polluted the Spanish coast with oil, and subsequently the European Union (EU) banned single-hull tankers carrying heavy oil from all EU ports [18].

### 8.7.2   ESTONIA ACCIDENT

This accident is concerned with an Estonian-flagged roll-on-roll (RO-RO) passenger ferry known as the *Estonia*. The *Estonia* left Tallinn, *Estonia*'s capital city, carrying 989 persons on board for Stockholm, Sweden, on September 27, 1994, and sank in the northern Baltic Sea in the early hours of September 28, 1994 [18]. The tragedy resulted in 852 deaths.

A subsequent investigation into the accident clearly revealed that the bow visor locks were too weak because of their poor manufacture and design. During bad weather, all these locks broke, and the visor fell off by pulling open the inner bow ramp [18,20].

### 8.7.3   HERALD OF FREE ENTERPRISE ACCIDENT

This accident is concerned with a passenger ship known as the Herald of Free Enterprise. The ship left Zeebrugge Harbour, Belgium, on March 6, 1987, and only 4 minutes after departure, it capsized and resulted in at least 150 passenger and 38 crew member deaths [18, 21]. The capsizing of the ship was due to a combination of adverse factors, including the vessel speed, the bow door being left open, and the trim by the bow.

The public inquiry into the Herald of Free Enterprise disaster was a very important milestone in ship safety in the United Kingdom. It resulted in actions such as changes to marine safety-associated rules and regulations, the development of a formal safety assessment process in the shipping industrial sector, and the introduction of the International Safety Management (ISM) Code for the safe operation of ships and for pollution prevention [18].

### 8.7.4 DERBYSHIRE ACCIDENT

This accident is concerned with a very large bulk carrier with a weight of 169,044 dwt (deadweight tons) called Derbyshire. The ship, which was en route to Kawasaki, Japan, carrying a cargo of iron ore concentrates, disappeared in puzzling circumstances during a typhoon in the Pacific on September 9, 1980 [18,20]. The tragedy resulted in 44 deaths (42 crew members and 2 wives).

The Derbyshire was designed in compliance with freeboard and hatch cover strengths as stated in the U.K. government's 1968 regulations [20]. The minimum hatch cover strength-related requirements for forward hatch covers for bulk carriers of similar size to the Derbyshire are considered quite seriously deficient in regard to the current acceptable safety levels [18].

## 8.8  SHIP PORT-ASSOCIATED HAZARDS

Over the years, there have been many ship port-associated accidents [22]. The ship port-associated hazards may be categorized under the following eight classifications [22]:

- **Classification I: Fire/explosion**. The hazards belonging to this classification are concerned with fire or explosion on the vessel or in the cargo bay. Three examples of these hazards are fire in the engine room, cargo tank fire/explosion, and fire in accommodation.
- **Classification II: Impacts and collision**. The hazards belonging to this classification are concerned with interaction with a stationery or a moving object, or a collision with a vessel. Three examples of these hazards are striking while at berth, vessel collision, and berthing impacts.
- **Classification III: Loss of containment**. The hazards belonging to this classification are concerned with the release and dispersion of dangerous substances. Two examples of these hazards are release of toxic material and release of flammables.
- **Classification IV: Navigation**. The hazards belonging to this classification are those that have potential for a deviation of the ship from its intended route or designated channel. Three examples of these hazards are pilot error, vessel not under command, and navigation error.
- **Classification V: Environmental**. The hazards belonging to this classification are those that take place when weather exceeds vessel design criteria

or harbour operation criteria. Three examples of these hazards are winds exceeding port criteria, extreme weather, and strong currents.

- **Classification VI: Manoeuvring**. The hazards belonging to this classification are concerned with failure to keep the vessel on the right track or to position the vessel as intended. Two examples of these hazards are berthing/unberthing error and fine-manoeuvring error.
- **Classification VII: Ship related**. The hazards belonging to this classification are concerned with ship-specific equipment or operations. Four examples of these hazards are loading/overloading, flooding, anchoring failure, and mooring failure.
- **Classification VIII: Pollution**. The hazards belonging to this classification are concerned with the release of material that can cause damage to the environment. Two examples of these hazards are crude oil spills and release of other cargo.

## 8.9   GLOBAL MARITIME DISTRESS SAFETY SYSTEM

Global Maritime Distress Safety System (GMDSS) is based upon a combination of satellite and terrestrial radio services and provides for automatic distress alerting and locating, thereby eliminating the need for a radio operator for sending an SOS/Mayday call (a Morse code distress signal). It may simply be characterized as an internationally agreed-upon set of safety-associated procedures, communication protocols, and equipment types employed for increasing safety and making it easier for rescuing distressed ships, aircraft, and boats.

This system (i.e., GMDSS) is composed of many systems. Some of them are new, and the others have been in operation for many years. GMDSS is intended to conduct the following main functions [23]:

- Maritime safety information broadcasts.
- Search and rescue coordination.
- Bridge-to-bridge communications.
- Alerting (including position determination of the unit in distress).
- General communications.
- Locating (homing).

Finally, it is to be noted that GMDSS also provides redundant means of distress alerting as well as emergency sources of power.

Some of the main types of equipment utilized in GMDSS are as follows [23]:

- **NAVTEX (Navigational Telex)**: This is an international automated system used for distributing maritime navigational-associated warnings, weather forecasts and warnings, search-and-rescue notices, etc.
- **Digital Selective Calling (DSC)**: This is a part of the GMDSS system and is basically intended for initiating ship-to-ship, ship-to-shore, and

shore-to-ship radiotelephone and medium frequency (MF)/ high frequency (HF) radiotelex calls.

- **Emergency Position-Indicating Radio Beacon (EPIRB)**: This equipment/system is designed for operating with the Cospas-Sarsat system, an international satellite-based search-and-rescue system established by the United States, Canada, France, and Russia. The automatically activated EPIRBs are designed for transmitting to alert rescue-coordination centres via the satellite system from any corner of the world.
- **Search-and-Rescue Transducer (SART) devices**: These devices are utilized for locating survival craft or distressed vessels by generating a series of dots on the rescuing ship's radar display.
- **Inmarsat Satellite Systems**: Inmarsat is a British satellite telecommunications company. The satellite systems operated by it and overseen by the International Mobile Satellite Organization (IMSO) are a very important element of the GMDSS. The Inmarsat ship-earth-station terminals B, C, and F77 are clearly recognized by the GMDSS.

## 8.10   SHIP SAFETY ASSESSMENT

Over the years, the assessment of ship safety has become an increasingly important issue. The methods/techniques of risk and cost-benefit assessments are utilized. The approach is called a formal safety assessment (FSA) and is composed of the following five steps [18]:

- **Step I: Identify hazards**: This step is concerned with the identification of hazards specific to a ship safety-associated problem under review. A hazard is defined as a physical situation with the potential for damage to property, human injury, damage to the surrounding environment, or some combination of these.
- **Step II: Assess risks**: This step is concerned with estimating risks and factors that influence the level of ship safety. The assessment of risks basically involves studying how hazardous events/states develop and interact for causing an accident/incident.
- **Step III: Propose risk-control options**: This step is concerned with proposing effective and practical risk-control options. The results of the previous steps are used for identifying high-risk areas in order to propose appropriate risk-control measures.
- **Step IV: Highlight benefits from reduced risks and costs**: This step is concerned with the identification of benefits from reduced risks and costs associated with the implementation of each of the previously highlighted risk-control options.
- **Step V: Make decisions**: This step is concerned with making decisions as well as providing appropriate recommendations for ship safety-associated improvements.

## 8.11   PROBLEMS

1. What are the causes of airplane crashes?
2. Write an essay on ship and airline safety.
3. What are the main responsibilities of the National Transportation Safety Board?
4. Discuss the United States airline-associated fatalities and accident rate.
5. Describe the following two data systems:
   - Accident/Incident reporting system.
   - Aviation safety reporting system.
6. List at least 12 data systems that can be useful for obtaining aviation safety-related information.
7. Describe the following two marine accidents:
   - Prestige accident.
   - Derbyshire accident.
8. Discuss ship port-associated hazards.
9. What are the main responsibilities of the Federal Aviation Administration?
10. Describe Global Maritime Distress Safety System.

## REFERENCES

1. Dhillon, B.S., *Human Reliability and Error in Transportation Systems*, Springer-Verlag, London, 2007.
2. ATAG, *Facts and Figures*, Air Transport Action Group, Geneva, Switzerland, retrieved on June 2008, from website: http://www.atag.org/content/showfacts.asp
3. IATA. 2006. *Fast Facts: The Air Transport Industry in Europe has United to Present Its Key Facts and Figures*, International Air Transport Association, Montreal, Canada, retrieved on December 2007, from website: http://www.iata.org/pressroom/economics -facts/stats/2003-04-10-01.htm
4. Dhillon, B.S., *Transportation Systems Reliability and Safety*, CRC Press, Boca Raton, Florida, 2011.
5. Bone, K., *The New York Waterfront: Evolution and Building Culture of the Port and Harbour*, Monacelli Press, New York, 1997.
6. Gardiner, R., *The Shipping Revolution: The Modern Merchant Ship*, Conway Maritime Press, London, 1992.
7. Wells, A.T., Rodrigues, C.C., *Commercial Aviation Safety*, McGraw Hill, New York, 2004.
8. Benowski, K., Safety in the Skies, *Quality Progress*, January 1997, pp. 25–35.
9. Howells, R., *The Myth of the Titanic*, Macmillan, Basingstoke, U.K., 1999.
10. Aviation Accident Synopses and Statistics, National Transportation Safety Board (NTSB), Washington, D.C., 2007, retrieved from website: http://www.ntsb.gov/Aviation/ Aviation.htm.
11. Zammit-Mangion, D., Eshelby, M., The Role of Take-Off Performance Monitoring in Future Integrated Aircraft Safety Systems, *IEE Digest*, No. 10203, 2003, pp. 2/1-2/9.
12. Matthews, S., *Safety Statistics: What the Number Tells Us About Aviation Safety at the End of the 20th Century*, Flight Safety Digest, December 1997, pp. 1–10.
13. Phillips, E.H., Global Teamwork Called Key to Improving Aviation Safety, *Aviation Weak and Space Technology*, July 2001, pp. 70–74.

14. Beck, H., Hosenball, M., Hager,M., Springen, K., Rogers, P., Underwood, A., Glick, D., Stanger, T., How Safe is This Flight ? *Newsweek*, April 24 1995, pp. 18–29.

15. Air Transport Association of America, *The Air Handbook*, Air Transport Association of America, Washington, D.C., 1995, retrieved from website: http://www.air-transport. org.

16. International Civil Aviation Organization, *Accident/Incident Reporting Manual (ADREP)*, Report No. 9156, International Civil Aviation Organization (ICAO), Montreal, Canada, 1987.

17. Boeing, 2001, *Statistical Summary of Commercial Jet Aircraft Accidents: Worldwide Operations 1959-2001*, Boeing Commercial Airplane Co., Seattle, Washington.

18. Wang, J., Maritime Risk Assessment and Its Current Status, *Quality and Reliability Engineering International*, Vol. 22, 2006, pp. 3–19.

19. The Prestige Casualty, Information Update No. 5, ABS press release, American Bureau of Shipping (ABS), Houston, Texas, 2002.

20. Wang, J., A Brief Review of Marine and Offshore Safety Assessment, *Marine Technology*, Vol. 39, No. 2, 2002, pp. 77–85.

21. United Kingdom Department of Transport, *Herald of Free Enterprise: Fatal Accident Investigation*, Report No. 8074, United Kingdom Department for Transport, Her Majesty's Stationery Office (HMSO), London, 1987.

22. Trbojevic, V.M., Carr, B.J., Risk Based Methodology for Safety Improvements in Ports, *Journal of Hazardous Materials*, Vol. 71, 2000, pp. 467–480.

23. Global Maritime Distress Safety System (GMDSS), *Wikipedia*, 2010, retrieved from website: http://en.wikipedia.org/wiki/Global_Maritime_Distress_Safety_System

# 9 Rail Safety

## 9.1 INTRODUCTION

Rail is a very important mode of transportation around the globe. Each year, millions of passengers and billions of dollars worth of goods are transported from one point to another through railroads.

Over the years, railway safety has been an important issue and in the United States, the Congress passed the Federal Railway Safety Appliances Act in 1893. The act instituted mandatory requirements for air brake systems and automatic couplers, as well as standardization of the locations and specifications for appliances. Over the years, due to actions such as this, the safety of rails throughout the United States has improved dramatically.

For example, as per FRA 2000, a Federal Rail Administration (FRA) report, the period 1993–1999 was the safest in U.S. rail-related history [1,2]. More specifically, during this period, train-accident-associated deaths dropped by 87%, rail worker casualties fell by around 34%, and rail/highway grade crossing-associated deaths decreased by more than 35% [1,2]. The rail industrial sector's US$50+-billion investment in infrastructure and equipment over the preceding decade is considered to be an important factor in reducing rail-related accidents.

This chapter presents various important aspects of rail safety.

## 9.2 CAUSES OF RAILWAY-ASSOCIATED ACCIDENTS AND INCIDENTS AND EXAMPLES OF THE CAUSES OF SPECIFIC RAIL ACCIDENTS

Over the years, there have been many different causes for the occurrence of railway-associated accidents and incidents around the globe. A study of 666 railway-associated accidents and incidents in Sweden during the period 1888–2000 grouped the causes for their occurrence under the following three classifications (along with their corresponding occurrence percentages in parentheses) [3]:

- **Classification I: Rail and track (39%)**: The causes under this classification are caused by or along the railway line, including the ballast, switches, sleepers, and objects placed on or quite close to the track. The classification also includes work on the track (e.g., maintenance and shunter actions).
- **Classification II: Rolling Stock (47%)**: The causes under this classification are related with track-bound vehicles such as trolleys and trains and include maintenance and operator-related errors.

DOI: 10.1201/9781003212928-9

- **Classification III: Insufficient information (14%)**: Accidents and incidents under this classification had insufficient information on accidents' and incidents' causes.

The study grouped the causes of 256 rail- and track-associated accidents under the following four categories (along with their corresponding occurrence percentages in parentheses) [3]:

- Category I: Railway operation (30%).
- Category II: Maintenance (30%).
- Category III: Sabotage (27%).
- Category IV: Uncertain (13%).

The maintenance-associated category was further divided into two groups (along with the corresponding occurrence percentage in parentheses): lack of maintenance (6%), and maintenance execution (24%).

Nine examples of the causes of some specific rail accidents are as follows:

- **Example I**: A Union Pacific Railroad train failed to stop at a signal and collided with another train in Macdona, Texas, United States, on June 28, 2004, and resulted in 3 deaths and 51 injuries [4].
- **Example II**: An Amtrak Auto-train derailed due to malfunctioning brakes and poor track maintenance near Crescent City, Florida, United States, on April 18, 2002, and resulted in 4 deaths and 142 injuries [5].
- **Example III**: A northbound Main South Line express freight train collided with a stationary southbound freight train due to a misunderstanding of track warrant conditions by both train drivers in Waipahi, New Zealand, on October 20, 1999, and resulted in one death and one serious injury [6].
- **Example IV**: A train derailed from a bridge damaged by road vehicles close to Gorey, Ireland, on December 31, 1975, and resulted in 5 deaths and 30 injuries [7].
- **Example V**: A passenger train, delayed by a cow on the line, was struck from behind by another passenger train wrongly signalled into section in Dundrum, Ireland, on December 23, 1957, and resulted in one death and four injuries [8,9].
- **Example VI**: A passenger train left the station without any train staff member and collided head-on with a freight train in Donegal, Ireland, on August 29, 1949, and resulted in three deaths and an unknown number of injuries [10].
- **Example VII**: A Cromwell to Dunedin passenger train derailed on a curve due to excessive speed because of an intoxicated driver in Hyde, New Zealand, on June 4, 1943, and resulted in 21 deaths and 47 injuries [6].
- **Example VIII**: A Wellington to Auckland express train rear-ended a northbound freight train after it passed a faulty semaphore signal that wrongly displayed clear instead of danger in Whangamarino, New Zealand, on May 27, 1914, and resulted in three deaths and five serious injuries [11].

- **Example IX**: A passenger train derailed due to excess speed on poor track in Ballinasloe, Ireland, on October 29, 1864, and resulted in 2 deaths and 34 injuries [10].

## 9.3 GENERAL CATEGORIES OF RAIL ACCIDENTS BY CAUSES AND EFFECTS

Over the years, various general categories of rail-related accidents according to causes and effects have been proposed. The common general categories of rail-related accidents by causes are as follows [2,12–14]:

- **Civil engineering failure**: This category includes bridge and tunnel collapses and track (i.e., permanent way) faults.
- **Signalmen's errors**: These include errors such as permitting two trains into the same already occupied block section and wrong operation of signals, points, or token equipment.
- **Contributory factors**: This category includes factors such as inadequate rules, rolling stock strength, effectiveness of brakes, and poor track or junction layout.
- **Drivers' errors**: These include errors such as excessive speed, engine mishandling, and passing signals at danger.
- **Acts of other people**: This category includes the acts of other railway personnel (e.g., porters, shunters) and of nonrailway personnel (i.e., vandalism, accidental damage, terrorism).
- **Mechanical failure of rolling stock** (because of poor maintenance and design).

Similarly, the commonly proposed general categories of rail-related accidents by effects are as follows [2,12–14].

- **Derailments**: These include plain track, junctions, and curves.
- **Collisions**: These include collisions with buffer stops, rear collisions, obstructions on the line/track (i.e., road vehicles, avalanches, landslides, etc.), and head-on collisions.
- **Others**: This category includes items such as falls from trains, collisions with people on tracks, and explosions and fires (including terrorism/ sabotage).

## 9.4 RAIL DERAILMENT ACCIDENTS AND INCIDENTS AND THEIR CAUSES

Over the years, there have been many rail derailment-related accidents and incidents around the globe, in which many deaths and injuries have occurred. Four examples of such accidents/incidents that occurred during the period 1833–2008 are as follows [2,15]:

- A train from Beijing, China, to Quingdao, China derailed in Shandong, China, on April 28, 2008, and resulted in 70 deaths and 400 injuries.
- A Pennsylvania Railroad express passenger train derailed near Altoona, Pennsylvania, United States, on February 18, 1947, and resulted in 24 deaths and 131 injuries.
- A military train derailed in Saint Michel de Maurienne, France, on December 12, 1917, and resulted in over 500 deaths. Up to the end of the 20th century, this disaster was considered the world's worst-ever derailment.
- A Camden and Amboy train derailed in Hightstown, New Jersey, United States, on November 11, 1833, and resulted in 2 deaths and 15 injuries.

There are many causes of rail derailment-related accidents and incidents. Some of the main ones are presented below [2,15]:

- **Wheel and truck failures**: Some of the main reasons for the occurrence of these failures are hot axle box, fracture of the wheel, and fracture of the axle.
- **Rail breakages**: These include rail breaks at rail joints, wheelburn-associated rail breaks, cold-weather-associated rail breaks, and manufacturing-defect-associated rail breaks.
- **In-train forces**: These occur due to factors such as train "stringlining" on sharp reverse curves, poor train-handling methods, and uneven loading.
- **Misaligned railroad tracks**: There are various types of misaligned plain line tracks that can contribute to or cause a derailment, including wide-to-gauge, washout, wrong alignment, incorrect cross-level, and wrong cant/superelevation.
- **Excessive speed**: There are two different mechanisms that cause excessive-speed derailments: rail roll and wheel climb. In the case of rail roll, the flange horizontal force applied to the gauge face of the rail is too high, overcoming the anchoring forces of rail clips and spikes. Similarly, in the case of wheel climb, the wheel is lifted off the track because the friction between the flange and gauge face of the rail is too high. Consequently, it causes the wheel flange to climb outwards over the rail's head.

## 9.5 TELESCOPING-ASSOCIATED RAILWAY ACCIDENTS

In rail accidents, telescoping takes place when the underframe of one vehicle overrides that of another vehicle and smashes through the body of the second vehicle. It is to be noted that the term "telescoping" is derived from the resulting appearance of the bodies of both vehicles. More clearly, the body of one vehicle may appear to be slid inside the body of the other vehicle like the tubes of a collapsible telescope (i.e., roof, underframe, and the body sides of the latter vehicle being forced apart from each other).

Past experiences clearly indicate that telescoping often resulted in heavy deaths when the telescoped train cars were fully occupied. The occurrence of telescoping-associated accidents can be lowered quite significantly with the use of anticlimbers and crash-energy-management structural systems.

Two important examples of rail telescoping-associated accidents are the Seer Green rail crash in the United Kingdom and the Chicago commuter rail crash in the United States. The Seer Green rail crash occurred on December 11, 1981, near Seer Green, Buckinghamshire, United Kingdom, when the driver of a train carrying passengers drove too fast for the surrounding conditions and ran into the back of an empty train at around 30 miles per hour [2,16]. The front coach of the train carrying passengers partly telescoped underneath the rear coach of the empty train and caused four deaths and five injuries. Additional information on the accident is available in Ref. [16].

The Chicago commuter rail crash, considered to be the worst in Chicago's history, took place on October 30, 1972, when Illinois Central Gulf Train 416 overshot the 27th Street station and collided with an express train. When the trains collided, the front car of the express train telescoped the rear car of the Illinois Central Gulf Train and caused 45 fatalities and 332 injuries [2,17]. Additional information on the accident is available in Ref. [17].

## 9.6 RAILWAY ACCIDENTS IN SELECTIVE COUNTRIES

Ever since the use of the steam engine for rail transportation, a vast number of accidents with fatalities, have occurred around the globe [18]. Railway-related accidents in three countries are presented in the following three subsections.

### 9.6.1 UNITED KINGDOM

Ever since the development of the steam engine by James Watt in the United Kingdom, there have been many railway accidents in the country (i.e., United Kingdom). The first railway passenger fatality took place on September 15, 1830, when William Huskisson was killed at the opening of the Liverpool and Manchester Railway line. In regard to fatalities, a railway accident that occurred on May 22, 1915, at Quintinshill was probably the worst rail accident in the United Kingdom. The accident caused 227 deaths and 246 injuries [18].

The approximate breakdowns of railway-associated number of accidents versus deaths occurring during the period 1830–2007 in the United Kingdom are presented in Table 9.1 [18].

### 9.6.2 AUSTRALIA

Over the years, there have been many rail-related accidents in Australia. During the period 1857–2007, there were about 41 major railway accidents. Australia's first fatal accident took place on June 25, 1857, when a rail worker fell from a locomotive as it was passing under a bridge [19,20].

In regard to fatalities, the worst railway accident in Australia, occurred on January 18, 1977, at Granville, New South Wales, when a packed peak-hour train derailed and crashed into a concrete bridge and caused 83 deaths. The breakdowns of major railway accidents that occurred during the period 1857–2007 in Australia are presented in Table 9.2 [19,20].

**TABLE 9.1**

**Approximate Breakdowns of Number of Railway Accidents versus Fatalities in the United Kingdom, 1830–2007**

| No. of accidents | No. of fatalities or fatality range |
|---|---|
| 15 | 1 |
| 7 | 2 |
| 4 | 3 |
| 7 | 4 |
| 11 | 5 |
| 23 | 6–9 |
| 76 | 10–50 |
| 5 | > 50 |

**TABLE 9.2**

**Breakdowns of Australia's Major Railway Accidents for the period 1857–2007**

| No. of accidents | Time period |
|---|---|
| 8 | 1857–1899 |
| 16 | 1900–1949 |
| 14 | 1950–1999 |
| 3 | 2000–2007 |

## 9.6.3 NEW ZEALAND

Since 1880, there have been many railway-related accidents in New Zealand. In fact as per Refs. [11,21], during the period 1880–2009, there were around 30 nonfatal/fatal rail-related accidents. The first reported fatal railway accident in New Zealand occurred on September 11, 1880, at the Rimutaka Incline, Wellington, and it resulted in three deaths and eleven injuries. The worst railway accident in the country took place on December 24, 1953, at Tangiwai, which caused 151 deaths [11].

The major railway accidents' breakdowns that occurred during the period 1880–2009 in New Zealand are presented in Table 9.3 [11,21].

## 9.7 LIGHT-RAIL TRANSIT SYSTEM SAFETY-ASSOCIATED ISSUES

Light-rail transit systems are being used in around 20 cities in Canada and the United States. Some of the main reasons for their usage are their relatively low costs and the ability for operating both on and off city streets, with intermediate capacity for transporting passengers and with frequent stops in urban areas.

**TABLE 9.3**
**Breakdowns of New Zealand's Major**
**Railway Accidents for the Period 1880–2009**

| No. of accidents | Time period |
| --- | --- |
| 2 | 1880–1889 |
| 9 | 1900–1949 |
| 17 | 1950–1999 |
| 2 | 2000–2009 |

Over the years, there have been many light-rail transit-system-associated accidents resulting in deaths and injuries. For example, during the three-year period from the opening of the Metro Blue Line (MBL) in Los Angeles, there were 24 train-pedestrian and 158 train-vehicle accidents/incidents resulting in 16 deaths and many injuries [22].

Nonetheless, some of the safety-related problems and areas of concern (i.e., safety-related issues) related to light-rail transit operations on city streets as well as on reserved rights of way with at-grade crossings are as follows [22]:

• Vehicle turning from roads or streets that run parallel to the rail tracks.
• Crossing equipment failure.
• Motorist confusion over light-rail transit signals, traffic signals, and signage at interaction points.
• Motorist disobedience in regard to traffic laws.
• Road vehicles making left or U-turns in front of rail vehicles or stopping on rail tracks.
• Pedestrian conflicts at crossing points and station areas.
• Light-rail vehicles blocking road/street and pedestrian crosswalk areas at crossing points.
• Traffic queues blocking crossing points.

## 9.8  RAILROAD TANK CAR SAFETY

Railroad tank cars are used for transporting liquids and gases from one point to another. Currently, in the United States alone, there are around 115,000 railroad tank cars in operation. These tank cars' contents are flammable, poisonous, corrosive, or pose other hazards if released accidentally. In the United States, during the period 1965–1980, more than 40 people were killed in tank car accidents [23]. Furthermore, accidental releases occur approximately once out of every 1,000 shipments, resulting in about 1,000 releases per year.

For ensuring tank car safety, the United States Department of Transportation and the industrial sector have taken various steps for enhancing both the tank car and the environment in which it functions. In the 1990 Hazardous Material Transportation Uniform Safety Act, the U.S. Congress called for the following two items [23]:

- An assessment of weather head shields should be made mandatory on all railroad tank cars that carry hazardous materials.
- An examination of the tank car design process.

In order to address these two items, the Transportation Research Board (TRB) formed a committee whose members had expertise in tank car design, transportation and hazardous materials safety, railroad operations and labour, transportation economics and regulation, chemical shipping, and chemical and mechanical engineering. After a careful examination of railroad tank car incident-related data, the committee recommended the following three measures [23]:

- **Measure I**: Improve the information and criteria employed for assessing the safety performance of tank car design types and for assigning materials to tank cars.
- **Measure II**: Enhance cooperation between industry and the Department of Transportation for identifying critical safety needs and taking action for achieving them.
- **Measure III**: Enhance the implementation of industry design approval and certification function and federal oversight procedures and processes.

## 9.9   METHODS FOR PERFORMING RAIL SAFETY ANALYSIS

There are many methods developed in areas such as reliability, safety, and quality for performing various types of analysis [24–26]. Some of these methods can also be used for performing rail safety analysis. These methods include fault tree analysis, interface safety analysis, Pareto diagram, hazards and operability analysis, failure modes and effect analysis, and cause-and-effect diagram. One of these methods (i.e., fault tree analysis) is presented below, and information on other methods is available in Chapter 4 and in Refs. [24–26].

### 9.9.1   Fault Tree Analysis

This is a widely used method for performing reliability and safety analysis of engineering systems in the industrial sector. The method was developed in the early 1960s at the Bell Telephone Laboratories for performing safety analysis of the Minuteman Launch control system, and it is described in Chapter 4 [25].

In order to demonstrate the application of fault tree analysis in the rail safety area, using the fault tree symbols defined in Chapter 4, a simple fault tree for the top event, release of liquefied chlorine from a rail tank shell, is shown in Figure 9.1 [24]. The capital letters in the rectangle and circles of the fault tree diagram in Figure 9.1

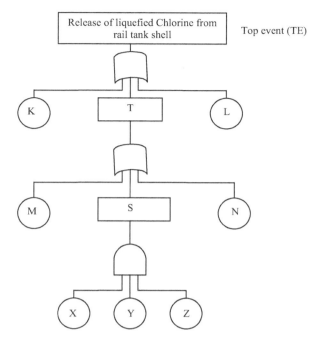

**FIGURE 9.1**   A fault tree for the top event: release of liquefied chlorine from a rail tank shell.

denote, respectively, intermediate and basic fault events associated with the rail tank shell. Each of these capital letters is defined below [24,25].

- X: Accident occurs.
- Y: Puncture load sufficient to fail tank head.
- Z: Puncture strikes tank head.
- S: Puncture probe fails tank head.
- M: Tank shell fails due to end impact.
- N: Crush load fails tank head.
- T: Release from tank head.
- K: Release from tank wall.
- L: Release from midway cover.

## Example 9.1

Assume that the occurrence probabilities of independent events X, Y, Z, M, N, K, and L in Figure 9.1 are 0.02, 0.03, 0.04, 0.05, 0.06, 0.07, and 0.08, respectively. Calculate the occurrence probability of the top event (i.e., release of liquefied chlorine from rail tank shell) with the aid of equations presented in Chapter 4.

Thus, from Chapter 4, the occurrence probability of event S is

$$P(S) = P(X)P(Y)P(Z) \tag{9.1}$$

where
P(X) is the occurrence probability of event X.
P(Y) is the occurrence probability of event Y.
P(Z) is the occurrence probability of event Z.

For the specified values of P(X), P(Y), and P(Z), from Equation (9.1), we obtain

$$P(S) = (0.02)(0.03)(0.04)$$

$$= 0.000024$$

Similarly, from Chapter 4, the occurrence probability of event T is

$$P(T) = 1 - [1 - P(M)][1 - P(S)][1 - P(N)] \tag{9.2}$$

where
P(M) is the occurrence probability of event M.
P(S) is the occurrence probability of event S.
P(N) is the occurrence probability of event N.

For the above calculated and the specified values of P(M), P(S), and P(N), Equation (9.2) yields

$$P(T) = 1 - [1 - 0.05][1 - 0.000024][1 - 0.06]$$

$$= 0.1070$$

Finally, the occurrence probability of the top event, release of liquefied chlorine from rail tank shell, is given by

$$P(TE) = 1 - [1 - P(K)][1 - P(T)][1 - P(L)] \tag{9.3}$$

where
P(TE) is the occurrence probability of the top event TE, the release of liquefied chlorine from a rail tank shell.
P(K) is the occurrence probability of event K.
P(L) is the occurrence probability of event L.

For the above calculated and specified values of P(T), P(K), and P(L), from Equation (9.3), we obtain

$$P(TE) = 1 - [1 - 0.07][1 - 0.1070][1 - 0.08]$$

$$= 0.2359$$

Thus, the occurrence probability of the top event, the release of liquefied chlorine from a tank shell, is 0.2359. Figure 9.1 fault tree, with the calculated and the given event occurrence probability values, is shown in Figure 9.2.

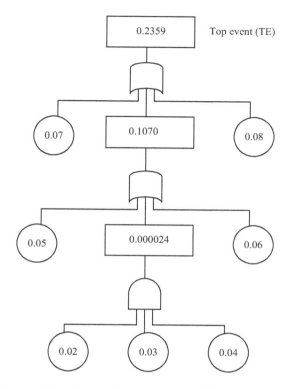

**FIGURE 9.2**  Redrawn Figure 9.1, fault tree with the calculated and the given event occurrence probability values.

## 9.10  PROBLEMS

1. Give at least four examples of rail-related accidents/incidents.
2. Write an essay on rail safety.
3. What are the common general categories of rail-related accidents by causes?
4. Discuss commonly proposed general categories of rail-related accidents by effects.
5. What are the main causes of rail derailment-related accidents and incidents?
6. Discuss telescoping-associated railway accidents.
7. Discuss railway accidents in the following three countries:
   (i)  New Zealand
   (ii)  United Kingdom
   (iii) Australia
8. List at least seven light-rail transit system safety-related issues.
9. Discuss railroad tank car safety.
10. List at least six methods that can be used to perform rail safety analysis.

# REFERENCES

1. Matthews, R.A., Partnerships Improve Rail's Safety Record, *Railway Age*, March 2001, pp. 14–15.
2. Dhillon, B.S., *Transportation Systems Reliability and Safety*, CRC Press, Boca Raton, Florida, 2011.
3. Holmgren, M., Maintenance-related Losses at the Swedish Rail, *Journal of Quality in Maintenance Engineering*, Vol. 11, No. 1, 2005, pp. 5–18.
4. *Macdona Accident*, 04–03, Report No. 04, National Transportation Safety Board, Washington, D.C., 2004.
5. *Derailment of Amtrak Auto Train P052-18 on the CSXT Railroad Near Crescent City, Florida*, April, 2002, Report No. RAR-03/02, National Transportation Safety Board, Washington, D.C., 2003.
6. Rail Accidents, Transportation Accident Investigation Commission, Wellington, New Zealand, 2000, retrieved from website: http://www.taic.org.nzl
7. Gorey Accident, Irish Railway Record Society (IRRS), Dublin, Ireland, 1976, retrieved from website: http://www.irrs.ie/
8. Murray, D., Collision at Dundrum, *Journal of the Irish Railway Record Society*, Vol. 17, No. 116, 1991, pp. 434–441.
9. MacAongusa, B., *The Harcourt Street Line: Back on Track*, Currach Press, Dublin, 2003.
10. MacAongusa, B., *Broken Rails*, Currach Press, Dublin, 2005.
11. Conly, G., Stewart, G., *Tragedy on the Track: Tangiwai and other New Zeland Railway Accidents*, Grantham House Publishing, Wellington, New Zealand, 1986.
12. Schneider, W., Mase, A., *Railway Accidents of Great Britain and Europe: Their Causes and Consequences*, David and Charles, Newton Abbot, U.K., 1970.
13. Rolt, L.T.C., *Red for Danger*, David and Charles, Newton Abbot, U.K., 1966.
14. Classification of Railway Accidents, *Wikipedia*, 2010, retrieved from website: http://en .wikipedia.org/wiki/classification_of_railway_accidents.
15. Derailment, *Wikipedia*, 2010, retrieved from website: http://en.wikipedia.org/wiki/ Derailment.
16. Rose, C.F., *Rail Accident: Report on the Collision that Occurred on December 11, 1981, Near Seer Green in the London Midland Region of British Railways*, Her Majesty's Stationery Office, London, 1983.
17. *Collision of Illinois Central Gulf Railroad Commuter Trains*, Report No. RAR-73-05, National Transportation Safety Board, Washington, D.C., 1973.
18. Rolt, L.T.C., Kichenside, G., *Red for Danger*, David and Charles, Newton Abbot, U.K., 1982.
19. Haine, E.A., *Railroad Wrecks*, Cornwall Books, New York, 1993.
20. Semmens, P.W.B., *Railway Disasters of the World*, Motor Books International, Minneapolis, Minnesota, 1994.
21. Churchman, G.B., Hurst, T., *Danger Ahead: New Zealand Railway Accidents in the Modern Era*, IPI Publishing Group, Wellington, New Zealand, 1992.
22. Meadow, L., Los Angeles Metro Blue Line Light-Rail Safety Issues, *Transportation Research Record*, No. 1433, 1994, pp. 123–133.
23. Ensuring Railroad Tank Car Safety, *TR News* 176, January–February 1995, pp. 30–31.
24. Dhillon, B.S., *Engineering Safety: Fundamentals, Techniques, and Applications*, World Scientific Publishing, River Edge, New Jersey, 2003.
25. Dhilon, B.S., *Design Reliability: Fundamentals and Applications*, CRC Press, Boca Raton, Florida, 1999.
26. Ishikawa, K., *Guide to Quality Control*, Asian Productivity Organization, Tokyo, 1976.

# 10 Truck and Bus Safety

## 10.1 INTRODUCTION

In most of the developed countries, commercial truck and bus transport have major economic importance. For example, in the United States, the commercial trucking industrial sector alone employs around 10 million personnel, and its annual revenue is higher than US$500 billion [1, 2]. Furthermore, charter and intercity buses in North America carry approximately 860 million passengers annually [1].

In 2003, there were 42,643 traffic crash deaths in the United States, out of which 4,986 involved large trucks. The crashes' economic impact involving large trucks and buses is quite significant. For example, in 2000, the crashes' average cost involving large trucks (i.e., greater than 10,000 lbs) was US$59,153, and the crashes' average cost involving transit or intercity bases was US$32,548 [3]. All in all, during the period 1997–1999, in the United States, the total annual costs of large truck crashes and bus crashes were around US$19.6 billion and US$0.7 billion, respectively [1].

This chapter presents various important aspects of truck and bus safety.

## 10.2 TOP TRUCK AND BUS SAFETY-RELATED ISSUES

Over the years, various studies have highlighted many truck and bus safety-associated issues. Some of the top ones are presented below [4].

- **Working conditions**: This is concerned with reviewing the standards and industry practices as they affect, directly or indirectly, driver workload.
- **Driver training**: This is concerned with the need for better and continuing education for all involved drivers (i.e., commercial and private).
- **Fatigue**: This is concerned with driving, unloading, scheduling, and road conditions that induce fatigue, in addition to hours-of-service violations and a lack of appropriate places to rest.
- **Communications**: This is concerned with the development of a national motor-carrier safety marketing-related campaign as well as the expansion of education efforts to the public for sharing roads with large vehicles.
- **Technology**: This is concerned with the development as well as deployment of emerging and practically inclined technologies for improving safety.
- **Accident countermeasures**: This is concerned with the research-related efforts targeted to seek and define proactive and non-punitive countermeasures for preventing accidents' occurrence.

DOI: 10.1201/9781003212928-10

- **Data/information**: This is concerned with the shortage of information, directly or indirectly, concerning heavy-vehicle crashes and their associated causes.
- **Uniform regulations**: This is concerned with the lack of uniformity in safety regulations and procedures among U.S. states as well as between Canada and Mexico, indicating that safety-related issues do not receive the same priority in all jurisdictions.
- **Partnership**: This is concerned with better communication and coordination among all highway users.
- **Resource allocations**: This is concerned with the priorities and allocation of scarce resources through better safety-management systems that clearly give safety the top priority.
- **License deficiencies**: This is concerned with the proper review of testing procedures for commercial driver's licenses.
- **Enforcement**: This is concerned with the need for clearly more effective testing and licensing, traffic enforcement, and adjudication of highway user violations.

## 10.3   TRUCK SAFETY-ASSOCIATED FACTS AND FIGURES

Some truck safety-associated facts and figures are as follows:

- In 2003, out of 4,986 deaths that resulted from crashes involving large trucks in the United States, 14% were occupants of large trucks, 78% were occupants of another vehicle, and 8% were no occupants [5,6].
- Large trucks account for approximately 3% of all registered vehicles in the United States and each truck on average travels around 26,000 miles per year [6].
- In 1993, as per the Federal Highway Administration's Office of Motor Carriers (OMC), approximately 80% of all truck accidents in the United States occurred with no adverse weather conditions [6].
- During the period 1976–1987, deaths of truck occupants in the United States decreased from 1,130 in 1976 to 852 in 1987 [7].
- In the United States, there were around 5,000, 4,900, 4,500, 4,000, 5,000, 5,100, and 5,400 truck-associated fatal crashes in 2000, 1997, 1995, 1992, 1989, 1986, and 1980, respectively [1].
- In 1993, around 4,500 trucks in the United States were involved in an accident in which at least one death occurred [6].
- As per the Centers for Disease Control (CDC), commercial drivers in the United States experience more job-associated deaths than any other profession [1,8].
- According to the Insurance Institute for Highway Safety around 65% of large truck crash deaths occur on major roads in the United States [6].
- During the period 1993–2003, the fatal crash rate for large trucks in the United States declined by 20% [1].

## 10.4 MOST COMMONLY CITED TRUCK SAFETY-ASSOCIATED PROBLEMS, FACTORS ASSOCIATED WITH HEAVY-VEHICLE ACCIDENTS, AND SAFETY CULTURE IN THE TRUCKING INDUSTRIAL SECTOR

The most commonly cited truck safety-associated problems are as follows [6]:

- **Rear-end Collisions**: These account for approximately 18.4% of fatal truck involvements, 30.6% of injury involvements, and 26.9% of tow-away involvements.
- **Rollovers**: These account for approximately 13.3% of fatal truck involvements, 10.8% of injury involvements, and 8.6% of tow-away involvements.
- **Angle impact**: This accounts for approximately 32.5% of fatal truck involvements, 37.1% of injury involvements, and 37.7% of tow-away involvements.
- **Jackknifes**: These take place when a multi-unit vehicle (e.g., tractor trailer) folds up like a pocket knife. Jackknifes account for approximately 8.3% of fatal truck involvements, 5.5% of injury involvements, and 8.4% of tow-away involvements.
- **Head-on collisions**: These account for approximately 24% of fatal truck involvements, 1.6% of injury involvements, and 1.9% of tow-away involvements.
- **Sideswipes**: These account for approximately 4.1% of fatal truck involvements and 9.7% of tow-away involvements.

The three most frequently associated factors, directly or indirectly, with heavy-vehicle accidents are shown in Figure 10.1 [7]. Additional information on these factors is available in Ref. [7].

Over the years, in many industrial settings, it has been recognized that the existence of a strong safety-related culture has a positive impact on safety-related outcomes such as severity and accident frequency [9–11]. The main factors that can affect safety-related culture in the trucking industrial sector include top management commitment to safety, driver scheduling autonomy, driver safety-related training,

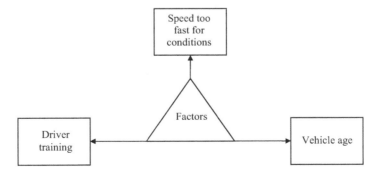

**FIGURE 10.1**    Most frequently associated factors with heavy-vehicle accidents.

and driver opportunities for safety input [12]. Careful consideration of such factors can quite significantly improve the safety-related culture in the trucking industrial sector, with a positive outcome on safety. Additional information on the safety-related culture in the trucking industrial sector is available in Ref. [12].

## 10.5   SAFETY-ASSOCIATED TRUCK INSPECTION TIPS, SAFETY-ASSOCIATED TIPS FOR TRUCK DRIVERS, AND RECOMMENDATIONS FOR IMPROVING TRUCK SAFETY

Past experiences over the years have clearly indicated that most truck-associated incidents result from faulty brakes and tires, electrical failures, mechanical failures, and cracked suspensions [6]. Detection is considered the most effective approach for avoiding such incidents' occurrence. This can easily be achieved with a daily "walk around" by drivers as well as through a regular preventive maintenance program. For example, during a trip, drivers should watch gauges for signs of trouble as well as use their senses (feel, look, smell, listen) to check for problems. Furthermore, whenever the truck is parked, the driver should properly check critical items such as wheels and rims, tires, electrical connections to the trailer, lights and reflectors, and brakes [6].

Some of the basic safety-associated tips for truck drivers are shown in Figure 10.2 [6].

In 1995, the attendees of a conference (Truck Safety: Perceptions and Reality) made many recommendations on five issues, shown in Figure 10.3, for improving truck safety.

The recommendations concerning the driver-fatigue issue were as follows [13]:

- Develop appropriate tolerance levels for accident risk and fatigue and devise all-new standards that clearly incorporate these levels.
- Develop a comprehensive approach for identifying the incidence of truck drivers' fatigue that clearly takes into consideration different types of fatigue and driving needs.
- Aim to harmonize all appropriate standards across different jurisdictions.

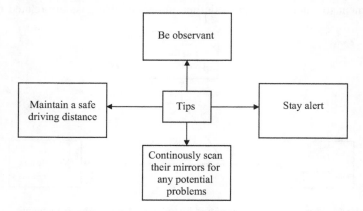

**FIGURE 10.2**   Basic safety-associated tips for truck drivers.

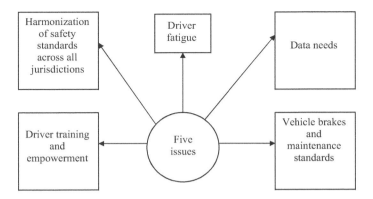

**FIGURE 10.3**  Five recommendation-associated issues for improving truck safety.

The recommendations concerning the driver training and empowerment issue were as follows [13]:

- Aim to devise and enforce regulations for ensuring that truck drivers are not unfairly dismissed for their refusal to drive in unsafe conditions.
- Develop effective driver training and retaining programs that specifically focus on safety (e.g., teaching drivers to inspect the vehicle by using the latest technology) and to take appropriate measures for reducing all types of accident risk.
- Aim to implement a graduated licensing scheme that appropriately reflects the needs of different types of trucking vehicles.
- Aim to enact a thorough accreditation of all driver training schools for ensuring that they uniformly satisfy desired standards in all jurisdictions in questions.

The recommendations concerning the vehicle brakes and maintenance standards issue were as follows [13]:

- Equip all trucks with appropriate on-board devices/signals for indicating when brakes need adjustment and servicing.
- Aim to train and certify all truck drivers for adjusting vehicle brakes appropriately as part of their ongoing training program and licensing requirements.
- Invoke appropriate penalties for all those trucking companies that regularly fail to satisfy all necessary inspection standards.
- Implement an appropriate safety rating system.

The recommendations concerning the data needs issue were as follows [13]:

- Improve police accident report reliability through better police training for collecting and reporting reliable data on accident-associated causes and consequences.

- Highlight and share currently available data concerning truck accidents and exposures.
- Establish an appropriate North American Truck Safety data centre.
- Aim to standardize police accident-related reporting forms in all concerned jurisdictions.

Finally, the recommendations concerning the harmonization of safety standards across all jurisdictions issue were as follows [13]:

- Form a committee of government and industry representatives for exploring appropriate avenues for cooperative efforts in developing uniform safety-related standards for trucks.
- Establish an agency to gather and disseminate safety-related information to all concerned parties.

## 10.6 BUS AND COACH OCCUPANT FATALITIES AND SERIOUS INJURIES

Buses and coaches are widely used for transporting passengers from one point to another around the globe. Each year, many passengers are killed and injured in bus- and coach-associated accidents. For example, in the United Kingdom alone, bus and coach occupant fatalities in 1991, 1990, 1985, 1980, 1975, 1970, and 1966 were 25, 19, 32, 29, 115, 74, and 76, respectively [14]. Furthermore, bus and coach passengers killed and seriously injured in all those same years were 725, 826, 1,036, 1,952, 1,650, 1,924, and 2,161, respectively.

Additional information on this topic is available in Ref. [14].

## 10.7 TRANSIT-BUS SAFETY AND KEY DESIGN-ASSOCIATED SAFETY FEATURE AREAS

Although buses are considered one of the safest modes of transportation, during the period 1999–2003 in the United States, an average of 40 bus occupant deaths and 18,430 injuries occurred per year [15]. In regard to two-vehicle crashes, there were 11 bus occupant deaths per year, while there were 162 deaths for occupants of other vehicles per year (i.e., 102 occupants in passenger cars, 49 in light trucks, 2 in large trucks, and 9 on motorcycles) [15].

During the period 1999–2001, in the United States, there was an average of 111 transit buses per year involved in fatal accidents [15,16]. Additional information on transit-bus safety is available in Refs. [15, 17].

Over the years, many design-associated feature areas for improving transit-bus safety have been identified. Four key safety-associated feature areas are as follows [18]:

- **Feature area I**: Interior transit-bus designs based on important considerations that clearly feature selective padding and removal of dangerous protrusions.

- **Feature area II**: Better all external designs that remove all types of potentially dangerous protrusions, handholds, and footholds.
- **Feature area III**: Wide doors, low floors, and energy-absorbing sidewalls and bumpers.
- **Feature area IV**: Better lighting and visibility for both driver and passengers.

## 10.8  VEHICLE SAFETY DATA SOURCES

There are many sources in the United States that can be used, directly or indirectly, for obtaining truck and bus safety-associated data. The main ones are as follows [19]:

- Federal Highway Administration.
- National Transportation Safety Board.
- National Highway Traffic Safety Administration.
- Insurance industry.

Each of these four sources is described below.

### 10.8.1  FEDERAL HIGHWAY ADMINISTRATION

Since 1973, the Federal Highway Administration (FHWA) has maintained a motor-carrier accident database, known as the Motor Carrier Safety Management Information System (MCMIS) [19]. It includes any federally regulated motor-carrier accident that meets the stated reporting criteria. Prior to 1986, the criteria demanded reporting of accidents resulting in a death, an injury, or property damage of US$2,000 or more.

The value of the property damage, in 1986 was increased to US$4,200 and in March 1987 to US$4,400. Subsequently, its value increased as per the Gross National Product index of inflation [19].

In comparison to any other national accident database, this database provides detailed truck accident-related characteristics. It includes information on items such as accident consequences, incident location, carrier identification and address, information on the cargo, characteristics of the event, and contributing factors.

The main strengths of the MCMIS are its exclusive truck focus and good detail on truck accident-related characteristics. In contrast, its main weaknesses are that it misses several portions of the truck population, its dependency on carrier participation, and concerns over the accuracy and its reports' completeness [19].

### 10.8.2  NATIONAL TRANSPORTATION SAFETY BOARD (NTSB)

This NTSB conducts multimodel, on-scene investigations of transportation-associated accidents. The basis for its jurisdiction for conducting an investigation is the definition of a major vehicular accident for each mode, as stated in the Code of Federal Regulations, part 49 [19].

In the latter years of the 1980s, the NTSB embarked on an extended study of heavy-truck safety [20] that included around 200 accidents involving heavy trucks meeting the following two conditions [19]:

- **Condition I**: The truck was damaged to a degree where it needed towing from the accident scene.
- **Condition II**: The accident involved a truck with gross-vehicle-weight rating higher than 10,000 lbs.

This NTSB study's strengths are the comprehensiveness of the accident investigation, the good detail on truck characteristics, and its exclusive focus on trucks. In contrast, its main weakness is the accidents' limited sample under investigation, which are not representative of truck crashes in general [19].

### 10.8.3 NATIONAL HIGHWAY TRAFFIC SAFETY ADMINISTRATION (NHTSA)

NHTSA's National Center for Statistics Analysis keeps records of various types of police-reported accidents [19]. The file of all reported accidents is referred to as the National Accident Sampling System (NASS), and it was started in the late 1970s. An accident to be included in NASS must meet the following three conditions [19,20]:

- **Condition I**: It must involve a motor vehicle in transport on a traffic way.
- **Condition II**: It must cause property damage/personal injury.
- **Condition III**: It must be reported by police.

The investigation team of NASS examines the accident scene and vehicle, reviews driver and medical records, and interviews all vehicle occupants. Each year, over 10,000 cases are investigated by various NASS teams and some of the main strengths of the NASS are as follows [19]:

- Comprehensive accident investigation.
- National estimate of accident frequency.
- Quite reasonable detail on truck accident characteristics.

In contrast, the NASS's two main weaknesses are the small number of heavy-truck accidents in the database and the lack of detailed analysis of the causes of accidents.

### 10.8.4 INSURANCE INDUSTRY

The U.S. insurance companies that underwrite motor carriers keep statistical and financial-associated information/data on insurance policies and claims. This information/data is also transmitted by all the participating companies to the Insurance Services Office, Inc. (ISO) [19, 21]. ISO is a non-profit organization that provides various types of data-associated services to U.S. property/casualty insurers [19,22].

**TABLE 10.1**
**Traffic-Accident-Associated Deaths and Deaths per**
**100,000 Vehicles in Selective Countries, 1996**

| No. | Country | Total deaths | Deaths per 100,000 vehicles |
|-----|---------|--------------|------------------------------|
| 1 | Ireland | 453 | 40.4 |
| 2 | Austria | 1,027 | 25.6 |
| 3 | Holland | 1,180 | 18.5 |
| 4 | Australia | 1,973 | 17.6 |
| 5 | Canada | 3,082 | 18.5 |
| 6 | Saudi Arabia | 3,123 | 62.4 |
| 7 | United Kingdom | 3,598 | 12.8 |
| 8 | France | 8,541 | 27.8 |
| 9 | Germany | 8,758 | 20.2 |
| 10 | Japan | 11,674 | 17.4 |
| 11 | Thailand | 15,000 | 125 |
| 12 | United States | 41,907 | 20.8 |

Its statistical data allow investigation of various industry characteristics, including driver age, geographic location, vehicle weight, vehicle age, and size of claim.

Finally, it is to be noted that this database has not been a primary source of information in carrying out safety analysis. Additional information on this topic is available in Refs. [19,21].

## 10.9 MOTOR VEHICLE TRAFFIC-ASSOCIATED ACCIDENTS IN SELECTIVE COUNTRIES

Motor vehicle traffic-associated accidents (including bus and truck) are becoming an important issue around the globe. Table 10.1 presents data on traffic-accident-associated deaths and deaths per 100,000 vehicles in selective countries for 1996 [22]. This clearly shows the importance of increasing the safety standards for both buses and trucks.

## 10.10 PROBLEMS

1. Discuss at least eight top bus and truck safety-associated issues.
2. What are the five most important truck safety-associated facts and figures?
3. Discuss at least five most commonly cited truck safety-associated problems.
4. What are the most frequently associated factors with heavy-vehicle accidents?
5. What are the basic safety-associated tips for truck drivers?
6. Discuss useful recommendations for improving truck safety.

7. Discuss the trends in deaths and serious injuries that have occurred in the United Kingdom for bus and coach occupants during the period 1966–1991.
8. Discuss transit-bus safety and key design-associated safety feature areas.
9. Describe the following two-vehicle data sources:
   • National Transportation Safety Board.
   • National Highway Traffic Safety Administration.
10. Write an essay on bus and truck safety.

## REFERENCES

1. The Domain of Truck and Bus Safety Research, Transportation research circular No. E-C117, Transportation Research Board (TRB), Washington, D.C., 2007.
2. Dhillon, B.S., *Transportation Systems Reliability and Safety*, CRC Press, Boca Raton, Florida, 2011.
3. Zaloshnja, E., Miller, T., *Revised Costs of Large Truck-and Bus-Involved Crashes*, Final report, Contract No. DTMC75-01-P-00046, Federal Motor Carrier Safety Administration (FMCSA), Washington, D.C., 2002.
4. Hamilton, S., The Top Truck and Bus Safety Issues, *Public Roads*, Vol. 59, No. 1, 1995, pp. 20.
5. *Large Truck Crash Facts*, Report No. FMCSA-RI-04-033, Federal Motor Carrier Safety Administration (FMCSA), Washington, D.C., 2005.
6. Cox, J., Tips on Truck Transportation, *American Papermaker*, Vol. 59, No. 3, 1996, pp. 50–53.
7. Seiff, H.E., Status Report on Large-Truck Safety, *Transportation Quarterly*, Vol. 44, No. 1, 1990, pp. 37–50.
8. Work-Related Road Crashes: Who's at Risk? Work-place Safety and Health Fact Sheet, DHHS (NOISH) publication No. 2004-137, Department of Health and Human Services (DHHS), Centers for Disease Control (CDC), Washington, D.C., 2004.
9. O'Toole, M.F., Successful Safety Committees: Participation Not Legislation, *Journal of Safety Research*, Vol. 33, 2002, pp. 231–243.
10. Gillen, M., Baltz, D., Gassel, M., Kirsch, L., Vaccaro, D., Perceived Safety Climate, Job Demands, and Co-worker Support Among Union and Non Union Injured Construction Workers, *Journal of Safety Research*, Vol. 33, 2002, pp. 33–51.
11. Zohr, D., Safety Climate in Industrial Organizations: Theoretical and Applied Implications, *Journal of Applied Psychology*, Vol. 65, No. 1, 1980, pp. 96–102.
12. Arboleda, A., Morrow, P.C., Crum, M.R., Shelley II, M. C., Management Practices as Antecedents of Safety Culture within the Trucking Industry: Similarities and Differences by Hierarchical Level, *Journal of Safety Research*, Vol. 34, 2003, pp. 189–197.
13. Saccomanno, F.F., Craig, L., Shortreed, J.H., Truck Safety Issues and Recommendations: Result of the Conference on Truck Safety: Perceptions and Reality, *Canadian Journal of Civil Engineers*, Vol. 24, 1997, pp. 326–332.
14. White, P., Dennis, N., Analysis of Recent Trends in Bus and Coach Safety in Britain, *Safety Science*, Vol. 19, 1995, pp. 99–107.
15. Olivares, G., Yadav, V., Mass Transit Bus-Vehicle Compatibility Evaluations During Frontal and Rear Collisions, Proceedings of the 20th International Conference on the Enhanced Safety of Vehicles, 2007, pp. 1–13.
16. Matteson, A., Blower, D., Hershberger, D., Woodrooffe, J., *Buses Involved in Fatal Accidents: Fact book 2001*, Center for National Truck and Bus Statistics, Transportation Research Institute, University of Michigan, Ann Arbor, Michigan, 2005.

17. Yang, C.Y.D., Trends in Transit Bus Accidents and Promising Collision Countermeasures, *Journal of Public Transportation*, Vol. 10, No. 3, 2007, pp. 217–225.
18. Mateyka, J.A., Maintainability and Safety of Transit Buses, Proceedings of the Annual Reliability and Maintainability Symposium, 1974, pp. 217–225.
19. Abkowitz, M., Availability and Quality of Data for Assessing Heavy Truck Safety, *Transportation Quarterly*, Vol. 44, No. 2, 1990, pp. 203–230.
20. NHTSA, *National Accident Sampling System (NASS): Analytical User's Manual*, National Highway Traffic Safety Administration (NHTSA), Washington, D.C., 1985.
21. ISD, *Insurance Data: A Close Look, Insurance Series*, Insurance Services Office (ISO), Jersey City, New Jersey, 1987.
22. Al Mogbel, A., Saudi Arabian Ministry of Communications and Its Role in Improving Traffic Safety, *Institute of Transportation Engineers (ITE) Journal*, June 2000, pp. 46–50.

# 11 Mining Equipment Safety

## 11.1 INTRODUCTION

Each year a vast sum of money is spent around the globe to produce various types of equipment for use in the mining sector. Nowadays, the type of equipment used in the mining area has come a long way since man first used tools made of flint and bone for extracting ores from the Earth. Some examples of the type of equipment utilized in the mining sector are haul trucks, crushers, mine carts, hoist controllers, and dragline excavators.

Over the years, many mine accidents have occurred that, directly or indirectly, involved mining equipment. In 1977, the U.S. Congress passed the Mine Safety and Health Act to improve safety in U.S. mines (including mining equipment safety). As the result of this Act, Mine Safety and Health Administration (MSHA) was established by the U.S. Department of Labour. The main goal of MSHA is to promote better health and safety conditions in the mining sector, enforce proper compliance with mine safety and health standards, and reduce health-associated hazards [1,2].

This chapter presents various important aspects of mining equipment safety.

## 11.2 MINING EQUIPMENT SAFETY-RELATED FACTS AND FIGURES

Some of the, directly or indirectly, mining equipment safety-related facts and figures are as follows:

- During the period 1978–1988, maintenance activities in the mines accounted for around 34% of all lost-time injuries [3].
- During the period 1995–2005, 483 fatalities in the U.S. mining operations were associated with equipment [1,2].
- During the period 1990–1999, electricity was the fourth leading cause for the occurrence of fatalities in the mining industrial sector [4].
- A study carried out by the U.S. Bureau of Mines (now National Institute for Occupational Safety and Health (NIOSH)) revealed that equipment was the basic cause of injury in around 11% of mining accidents and a secondary causal factor in the occurrence of another 10% of the accidents [5–7].
- In 2004, around 17% of the 37,445 injuries in the underground coal mines were connected, directly or indirectly, to bolting machines [8].
- During the period 1983–1990, around 20% of the coal mine-related injuries occurred during the equipment maintenance activity or while using hand-held tools [9].
- During the period 1990–1999, 197 equipment fires in the coal mining operations caused 76 injuries [10].

DOI: 10.1201/9781003212928-11

## 11.3   TYPES OF MINING EQUIPMENT INVOLVED IN FATAL ACCIDENTS AND THE FATAL ACCIDENTS' BREAKDOWNS AND MAIN CAUSES OF MINING EQUIPMENT ACCIDENTS

Over the years, many fatal accidents involving various types of equipment used in mines, have occurred. For example, an MSHA study reported that, during the period 1995–2005, there were 483 equipment-associated fatal accidents in U.S. mining operations [11]. The type of equipment involved is presented below and its corresponding fatal accidents, for the stated time period, are in parentheses [2,11]:

- Hoisting (2).
- Longwall (5).
- Forklift (5).
- LHD (load-haul-dump) (6).
- Roof bolter (7).
- Shuttle car (13).
- Drill (16).
- Dozer (28).
- Miner (30).
- Front-end loader (41).
- Conveyor (41).
- Haul truck (108).
- Miscellaneous equipment (177).

Furthermore, the percentage distributions of the above accidents with respect to the equipment type involved were 0.41% (hoisting), 1.04% (longwall), 1.04% (forklift), 1.24% (LHD), 1.45% (roof bolter), 2.69% (shuttle car), 3.31% (drill), 5.8% (dozer), 6.21% (miner), 8.49% (front-end loader), 9.32% (conveyor), 22.36% (haul truck), and 36.65% (miscellaneous equipment) [11]. It means over 50% of the fatal accidents were just due to five types of equipment (i.e., haul truck, conveyor, front-end loader, miner, and dozer).

Over the years, many studies have been conducted for highlighting main causes of mining equipment accidents. One such study was conducted by the U.S. Bureau of Mines (now NIOSH). The study highlighted the following seven main causes for the occurrence of mining equipment accidents [6]:

- Poor control display layout.
- Poor original design or redesign.
- Poor ingress/egress design.
- Unguarded moving parts.
- Restricted visibility.
- Hot surfaces/exposed wiring.
- Exposed sharp surfaces/pinch points.

## 11.4  MINING ASCENDING ELEVATOR ACCIDENTS AND FATALITIES AND INJURIES DUE TO DRILL RIG, HAUL TRUCK, AND CRANE CONTACT WITH HIGH-TENSION POWER LINES

Over the years many ascending elevator-related accidents in mining operations have occurred and a number of them resulted in deaths or serious injuries. All these accidents occurred on counterweighted elevators due to structural, mechanical, and electrical failures. Past experiences over the years indicate that, although the elevator cars have safe ties that grip the guide rails and stop a falling car, such devices fail for providing an appropriate level of protection in the upward direction.

In 1987, an ascending elevator car-related accident at a Pennsylvania coal mine caused extensive structural damage and disabled the elevator for many months [12]. As the result of this accident, in order to provide effective ascending car over speed protection for new and current installations, the Pennsylvania Bureau of Deep Mine Safety established an advisory committee for evaluating devices on the market. After an extensive investigation, the committee recommended four protective methods, shown in Figure 11.1, as well as requirements that all new elevators must have a governor rope monitoring device, a manual reset, and back out of over travel switch [2,12].

Additional information on the methods shown in Figure 11.1 is available in Refs. [12–14]. Past experiences over the years clearly indicate that overhead or high-tension power lines present a serious electrocution hazard to people employed in a variety of industrial sectors because equipment such as cranes, haul trucks, and drill rigs are often exposed to these power lines. When contacting with the power lines, this type

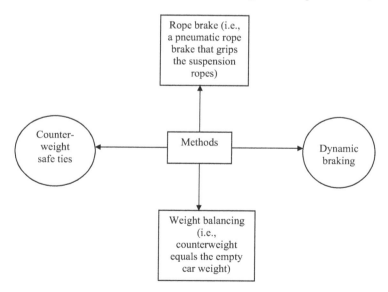

**FIGURE 11.1**  Mining ascending elevator protective methods.

of equipment becomes quite susceptible to a high voltage, and simultaneous contact to the "hot" frame and ground by people can result in electric shocks and dangerous burns. The industrial sectors where the risk of occurrence of such accidents is highest include mining, construction, and agriculture.

Each year in the United States approximately 2,300 accidental overhead power line contacts occur [15]. During the period 1980–1997, the U.S. mining industry reported at least 94 mobile equipment overhead line contact accidents [15]. These accidents caused 114 injuries and around 33% of them were fatal. Most of these accidents involved cranes (47%), dump bed trucks (24%), and drills (14%) [15].

## 11.5  PROGRAMMABLE ELECTRONIC-ASSOCIATED MINING MISHAPS AND LESSONS LEARNED

Nowadays, programmable electronic-associated mishaps have become an important issue in the mining industrial sector. The MSHA's serious thinking about the functional safety of programmable electronics (PE)-based mining systems began in 1990 because of an unplanned longwall shield mishap [16]. During the period 1995–2001, there were 11 PE-associated mining incidents in the United States; four of them resulted in deaths [17,18]. As per Ref. [18], during the same time period (i.e., 1995–2001), there were 71 PE-associated incidents in underground coal mines in New South Wales, Australia. A study of both these data sets clearly reported that most mishaps or incidents involved sudden movements or start-ups of PE-based mining systems.

In 1991, MSHA carried out a study of longwall installations in regard to programmable electronics. The study reported that approximately 35% had experienced sudden movements basically due to the following four problems:

1. Software programming errors.
2. Defective or sticking solenoid valves.
3. Water ingress.
4. Operator errors.

A detailed analysis of the Australian and U.S. data sets revealed that there were four factors (i.e., improper operation, solenoid valves, water ingress, software) that, directly or indirectly, contributed to PE-based mishaps [16,19]. However, the solenoid valve-associated problems were the main contributing factor to PE-based mishaps or incidents.

In response to the longwall-associated mishaps, MSHA recommended improvements in four areas: timely maintenance, operator training, maintaining alertness for abnormal operational sequences that might be indicative of a software problem, and maintaining integrity of enclosure sealing. Subsequently, in order to overcome the shortcomings of the above approach, MSHA proposed a safety framework largely based on the International Electrotechnical Commission (IEC) 61508 safety life cycle [20].

Over the years, many valuable lessons have been learned for addressing programmable electronic mining system safety-associated issues. Most of these lessons include [18]:

- Making use of appropriate scenarios for conveying information.
- Establishing appropriate terminology, definitions, and concepts as early as possible.
- Identifying and clearly understanding all associated issues and perceptions.
- Decomposing the involved problem into manageable parts.
- Involving all appropriate elements of the industrial sector as early as possible and on a continuous basis.
- Holding industry workshops as considered appropriate.
- Clearly separating all the associated concerns.

## 11.6   EQUIPMENT FIRE-ASSOCIATED MINING ACCIDENTS AND MINING EQUIPMENT FIRE IGNITION SOURCES

Over the years, there have been many equipment-associated mining accidents that have caused injuries. For example, during the period 1990–1999, in U.S. coal mines there were 197 equipment fires that resulted in 76 injuries [10]. For further study, NIOSH grouped all these fires under the following four classifications [10]:

- **Classification I**: Underground equipment fires. There were 26 equipment fires and 10 of them resulted in 10 injuries.
- **Classification II**: Surface coal mine equipment fires. There were 14 equipment fires and 4 of them resulted in 4 injuries.
- **Classification III**: Surface equipment fires at underground coal mines. There were 140 equipment fires and 56 of them resulted in 56 injuries.
- **Classification IV**: Prep plant fires. There were 17 equipment fires and 6 of them resulted in 6 injuries.

Over the years, many studies have been conducted to highlight ignition sources for mining equipment fires. The four main ones are as follows [10]:

- Electric short/arcing.
- Hydraulic fuel/fluid on equipment hot surfaces.
- Engine malfunction.
- Flame welding/cutting spark/slag.

Additional information on all these ignition sources is available in Ref. [10].

## 11.7   STRATEGIES FOR REDUCING MINING EQUIPMENT FIRES AND USEFUL GUIDELINES FOR IMPROVING ELECTRICAL SAFETY IN MINES

Over the years, many different ways have been explored for reducing mining equipment fires and injuries. Some of the better strategies/methods are as follows [10]:

- Conduct equipment hydraulic, fuel, and electrical system inspections frequently and thoroughly.

- Aim for developing new technologies for fire barriers, emergency engine/pump shutoff, and emergency hydraulic line drainage/safeguard system.
- Provide frequent and necessary emergency preparedness training to all involved equipment operators.
- Improve as necessary equipment/cab fire prevention/suppression systems.
- Develop effective equipment/cab fire detection systems that have an audible/visible cab alarm.

Over the years many guidelines for improving general electrical safety in the mining industrial sector (including equipment) have been proposed. Five of these guidelines considered most useful are as follows [4,21]:

1. Make improvements in system/equipment design in general and in regard to electrical safety in particular.
2. Make all necessary improvements in electrical maintenance procedures and schedules.
3. Provide appropriate power line awareness training to all concerned individuals.
4. Make use of appropriate power line avoidance devices as much as possible.
5. Target appropriate training in all problem areas.

## 11.8   HUMAN FACTORS-ASSOCIATED DESIGN TIPS FOR SAFER MINING EQUIPMENT

A number of studies carried out over the years have clearly indicated that equipment is the primary cause of injury in approximately 11% of mining accidents and a secondary cause in another 10%. Therefore, it is of utmost importance that all new mining equipment clearly incorporate good human-factors design criteria that maximize safety of all involved mine workers. In this regard, human factors-associated tips considered most useful are divided into four areas as shown in Figure 11.2 [5,21].

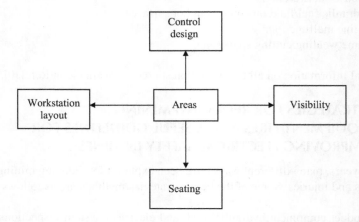

**FIGURE 11.2**   Areas of human factors-associated design tips for safer mining equipment.

Four tips concerning the workstation layout are as follows [5,21]:

1. Aim for distributing workload as evenly as possible between hands and feet.
2. Anticipate all types of potential safety-associated hazards and necessary emergency measures prior to starting the design process.
3. Ensure that the relative placement of displays and controls for similar equipment is maintained effectively.
4. Ensure that the workstation fits all operators effectively from the 5th to 95th percentile range.

Five tips concerning the control design area are as follows [5,21]:

1. Ensure that all design controls can clearly withstand or guard against abuse, such as from the forces imposed during a panic response in an emergency situation or from falling ribs and roof.
2. Ensure that all types of design controls properly comply with anthropometric data on human operators.
3. Ensure that all controls have adequate resistance for decreasing the possibility of inadvertent activation by the weight of a foot or hand.
4. Ensure that all involved operators can highlight the appropriate controls, quickly and accurately.
5. Ensure that a vehicle's or a part's speed is proportional to the control displacement from its rest position as well as in the same direction.

Two tips concerning the visibility area are as follows [5,21]:

1. Ensure that the workstation provides an unobstructed line of sight to locations/objects that should be clearly visible for performing a task effectively and safely.
2. Ensure that there is sufficient contrast between the luminance of the object or location of interest and the surrounding background so that the required task can be conducted safely and effectively.

Finally, five tips associated with the seating area are as follows [5,21,22]:

1. Ensure that the seat properly adjusts and fits to body dimensions, distributes weight for relieving pressure points, and supports posture.
2. Design the seat in such a way that mineworkers at large can easily maintain or replace it.
3. Ensure that the seat provides essential design features for guarding against shocks due to rough roads and minor collisions that tend to unseat the person occupying the seat.
4. Ensure that the seat does not hinder the ability of the involved operator for controlling the machine/equipment.
5. Ensure that the seat does not hinder the involved operator's ability to enter or exit the workstation.

## 11.9   METHODS FOR PERFORMING MINING
## EQUIPMENT SAFETY ANALYSIS

There are a large number of methods available in the published literature that can be used for performing mining equipment safety analysis [23]. Nine of these methods considered most useful for performing mining equipment safety analysis are as follows [23]:

1. Consequence analysis.
2. Management oversight and risk tree analysis.
3. Human reliability analysis (HRA).
4. Binary matrices.
5. Root cause analysis.
6. Failure modes and effect analysis (FMEA).
7. Hazards and operability analysis.
8. Preliminary hazards analysis.
9. Technic of operations review.

The first four of the above nine methods are presented below and the remaining five methods are described in Chapter 4.

### 11.9.1   Consequence Analysis

This method is concerned with determining the impact of an undesired event on items such as adjacent property, people, or the environment. The typical examples of an undesired event are fire, explosion, projection of debris, and the release of toxic material. Some of the primary consequences of concern in the area of mining are injuries, deaths, operational downtime, and losses due to property/equipment damage.

All in all, consequence analysis is safety analysis's one of the intermediate steps as accident consequences are generally determined, initially using methods such as preliminary hazards analysis or failure modes and effect analysis. Additional information on this method (i.e., consequence analysis) is available in Ref. [24] and information on its application in the mining area is available in Ref. [25].

### 11.9.2   Management Oversight and Risk Tree Analysis

This is a comprehensive safety assessment method that can be applied to any mine safety-related program and is based on a document prepared in 1973, by W.G. Johnson (director at the Atomic Energy Commission) [26]. The method focuses on administrative or programmatic control of hazardous conditions and is particularly designed for highlighting, evaluating, and preventing the occurrence of safety-related errors, omissions, and oversights by management and workers that can result in accidents.

The following nine steps are followed to conduct management oversight and risk tree (MORT) analysis [25]:

- **Step 1**: Obtain sufficient working knowledge of the equipment/system under study.
- **Step 2**: Select the accident for analysis.
- **Step 3**: Highlight potential hazardous energy flows and barriers related to the accident sequence.
- **Step 4**: Document all necessary information in the standard MORT-type analytical tree format.
- **Step 5**: Determine all possible factors that can, directly or indirectly, cause initial unwanted energy flow.
- **Step 6**: Document all the safety program elements that are considered to be less than sufficient in regard to the unwanted energy flow.
- **Step 7**: Continue conducting analysis of all the safety program elements in regard to the rest of the unwanted energy flows (if any).
- **Step 8**: Determine all the management system factors associated with the potential accident.
- **Step 9**: Review the accomplished analysis for all safety program elements that could reduce the likelihood of the potential accident's occurrence.

Some of the benefits and drawbacks of the MORT analysis method are as follows [26]:

### 11.9.2.1  Benefits
- Useful for evaluating all three aspects of an industrial system (i.e., management, hardware, and human) as they collectively cause accidents.
- A comprehensive and effective approach that attempts to review each and every aspect of safety in any type of work.
- Results of MORT analysis can suggest appropriate improvements to an existing safety program that could be quite helpful in reducing property damage, decreasing injuries, and saving lives.

### 11.9.2.2  Drawbacks
- Emphasizes management's responsibility for providing a safe work environment.
- A time-consuming approach/method.
- Creates a vast amount of complex detail.

### 11.9.3  Human Reliability Analysis (HRA)

Human reliability analysis may simply be described as human performance's study within the framework of a complex man–machine operating system. The method is considered very useful in developing qualitative information concerning human errors' causes and effects in specific situations. HRA is conducted by following the six steps presented below [25,27]:

- **Step 1**: Describe the goals and functions of the system under consideration.
- **Step 2**: Describe all possible situational characteristics.

- **Step 3**: Describe the characteristics of all the involved personnel.
- **Step 4**: Describe the tasks and jobs conducted by the involved personnel.
- **Step 5**: Conduct an analysis of the tasks and jobs for identifying error-likely situations and other likely problems.
- **Step 6:** Suggest all necessary changes to the system under consideration.

Additional information on the above six steps is available in Refs. [24,25]. The application of this method to a mining problem is demonstrated in Ref. [25].

### 11.9.4  BINARY MATRICES

This is a useful, logical, and qualitative method for identifying system interactions [28]. The method can be used during the safety analysis system-description stage, or as a final checkpoint in a preliminary hazards analysis or FMEA for ensuring that all important system dependencies have been considered in the analysis effectively.

The specific tool utilized in binary matrices is the binary matrix that contains information on the relationships between all system elements. The primary objective of binary matrix is to highlight the one-on-one dependencies that exist between all system elements. All in all, this matrix serves merely as a useful tool for "reminding" the analyst that failures in one part of a system/equipment may effect the normal operation of other subsystems in completely distinct areas.

Information on the application of this method to a mining system is available in Ref. [25].

## 11.10  HAZARDOUS AREA SIGNALLING AND RANGING DEVICE (HASARD) PROXIMITY WARNING SYSTEM

Over the years, many underground and surface mining workers working close to machinery and powered haulage have been permanently disabled or killed. For example, as per Ref. [22], each year in surface mining-related operations around 13 persons are killed by being run over or pinned by mobile equipment. Furthermore, in underground mining-related operations in the United States alone, during the period 1988–2000, 23 deaths were associated with mining workers getting crushed, caught, or pinned by continuous mining equipment/systems.

A subsequent analysis of all these deaths clearly revealed that, from time to time, mining workers become totally preoccupied with operating their own equipment and fail to realize when they stray into or are subjected to hazardous conditions.

For overcoming problems such as these, NIOSH developed an active proximity warning system known as hazardous area signalling and ranging device. Over the years, this system has repeatedly proved to be a quite useful tool for warning mineworkers when they approach hazardous areas around heavy mining equipment and other hazardous areas. The system is composed of the following two subsystems [21,22]:

- **Subsystem I: Transmitter**: It produces a 60 kHz magnetic field with the aid of one or more wire loop antennas. In turn, each antenna is adjusted for establishing a magnetic field pattern for each hazardous area, as the need arises.

- **Subsystem II**: **Receiver**: It is a magnetic field metre and is worn by the mining workers. It compares the received signal with preset levels, which are calibrated for highlighting dangerous levels. Furthermore, the receiver outputs can be made for stopping ongoing equipment operations and can include audible, visual, and vibratory indicators.

Additional information on this system is available in Ref. [22].

## 11.11   PROBLEMS

1. List at least nine types of mining equipment involved in fatal accidents.
2. List at least five important facts and figures, directly or indirectly, concerning mining equipment safety.
3. Discuss mining ascending elevator-related accidents.
4. What are the seven important lessons learned in addressing programmable electronic mining system safety-associated issues?
5. What are the four main mining equipment fire ignition sources?
6. What are the important strategies for reducing mining equipment fires?
7. Discuss human factors-associated design tips for safer mining equipment.
8. What are the benefits and drawbacks of management oversight and risk tree analysis?
9. List at least eight methods that can be used to conduct mining equipment safety analysis.
10. Describe hazardous area signalling and ranging device (HASARD) proximity warning system.

## REFERENCES

1. Mine Safety and Health Administration (MSHA), U.S. Department of Labor, Washington, D.C, retrieved from website http:www.msha.gov/.
2. Dhillon, B.S., *Mine Safety: A Modern Approach*, Springer-Verlag, London, 2010.
3. *MSHA Data for 1978–1988*, Mine Safety and Health Administration (MSHA), U.S. Department of Labor, Washington, D.C.
4. Cawley, J.C., Electrical Accidents in the Mining Industry: 1990–1999, *IEEE Transactions on Industrial Applications*, Vol. 39, No. 6, 2003, pp. 1570–1576.
5. Unger, R.L., Tips for Safer Mining Equipment, U.S. Department of Energy's Mining Health and Safety Update, Vol. 1, No. 2, 1996, pp. 14–15.
6. Saunders, M.S., Shaw, B.E., *Research to Determine the Contribution of System Factors in the Occurrence of Underground Injury Accidents*, Report No. USBM OFR 26-89, U.S. Bureau of Mines (USBM), Washington, D.C., 1988.
7. What Causes Equipment Accidents?, National Institute for Occupational Safety and Health (NIOSH), 2008, retrieved from website http://www.cdc.gov/niosh/mining/topics/machinesafety/equipmentdsgn/equipment accident
8. Burgess-Limerick, R., Steiner, L., Preventing Injuries: Analysis of Injuries Highlights High Priority Hazards Associated with Underground Coal Mining Equipment, *American Longwall Magazine*, August 2006, pp. 19–20.
9. Rethi, L.L., Barett, E.A., *A Summary of Injury Data for Independent Contractor Employees in the Mining Industry from 1983–1990*, Report No. USBM IC 9344, U.S. Bureau of Mines, Washington, D.C., 1983.

10. De Rosa, M., Equipment Fires Cause Injuries: Recent NIOSH Study Reveals Trends for Equipment Fires at U.S. Coal Mines, *Coal Age*, No. 10, 2004, pp. 28–31.
11. Kecojevic, V., Komljennovic, D., Groves, W., Rodomsky, M., An Analysis of Equipment-Related Fatal Accidents in U.S. Mining Operations: 1995–2005, *Safety Science*, Vol. 45, 2007, pp. 864–874.
12. Barkand, T.D., Ascending Elevator Accidents: Give the Miner a Brake, *IEEE Transactions on Industry Applications*, Vol. 28, No. 3, 1992, pp. 720–729.
13. Barkand, T.D., Helfrich, W.J., Application of Dynamic Braking to Mine Hoisting Systems, *IEEE Transactions on Industry Applications*, Vol. 24, No. 5, 1988, pp. 507–514.
14. Nederbragt, J.A., Rope Brake: As Precaution Against Overhead, *Elevator World*, Vol. 7, 1989, pp. 6–7.
15. Sacks, H.K., Cawley, J.C., Homce, G.T., Yenchek, M.R., Feasibility Study to Reduce Injuries and Fatalities Caused by Contact of Cranes, Drill Rigs, and Haul Trucks with High-Tension Lines, *IEEE Transactions on Industry Applications*, Vol. 37, No. 3, 2001, pp. 914–919.
16. Dransite, G.D., Ghosting of Electro-Hydraulic Long Wall Shield Advanced Systems, Proceedings of the 11th West Virginia University International Electro Technology Conference, 1992, pp. 77–78.
17. *Fatal Alert Bulletins, Fatal Grams and Fatal Investigation Reports*, Mine Safety and Health Administration (MSHA), Washington, D.C., May 2001, retrieved from website: http://www.msha.gov/fatals/fab.htm.
18. Sammarco, J.J., Addressing the Safety of Programmable Electronic Mining Systems: Lessons Learned, Proceedings of the 37th IEEE Industry Applications Society Meeting, 2003, pp. 692–698.
19. Waudby, J.F., *Underground Coal Mining Remote Control of Mining Equipment: Known Incidents of Unplanned Operation in New South Wales (NSW) Underground Coal Mines*, Dept. of Mineral Resources, NSW Department of Primary Industries, Maitland, NSW, Australia, 2001.
20. *IEC 61508, Parts 1–7, Functional Safety of Electrical/Electronic/Programmable Electronic Safety-Related Systems*, International Electrotechnical Commission (IEC), Geneva, Switzerland, 1998.
21. Dhillon, B.S., *Mining Equipment Reliability, Maintainability, and Safety*, Springer-Verlag, London, 2008.
22. NIOSH, *Hazardous Area Signalling and Ranging Device (HASARD)*, Pittsburgh Research Laboratory, National Institute for Occupational Safety and Health (NIOSH), Atlanta, GA.
23. Dhillon, B.S., *Engineering Safety: Fundamentals, Techniques, and Applications*, World Scientific Publishing, River Edge, NJ, 2003.
24. American Institute of Chemical Engineers, *Guidelines for Consequence Analysis of Chemical Releases*, American Institute of Chemical Engineers, New York, 1999.
25. Daling, P.M., Geffen, C.A., *User's Manual of Safety Assessment Methods for Mine Safety Officials*, Report No. BuMines OFR 195(2)-83, U.S. Bureau of Mines, Department of the Interior, Washington, D.C., 1983.
26. Johnson, W.G., *The Management of Oversight and Risk Tree-MORT*, Report No. SAN 821-2, U.S. Atomic Energy Commission, Washington, D.C., 1973.
27. Dhillon, B.S., *Safety and Human Error in Engineering Systems*, CRC Press, Boca Raton, Florida, 2013.

# 12 Programmable Electronic Mining System Safety

## 12.1 INTRODUCTION

Generally, mining was traditionally a low-tech industrial area. Nowadays, it is driven by competitive pressures to go high-tech by utilizing programmable electronics (PE) in areas such as longwall mining systems, mine processing equipment, automated haulage, and mine monitoring systems. Just as in other industrial areas, PE technology's (i.e., software, microprocessors, and programmable logic controllers) application in the mining area has created unique challenges for system design, verification, operation, maintenance, and assurance of functional safety [1,2]. More clearly, although PE's application provides numerous benefits, it certainly adds a level of complexity that, if not considered carefully, may compromise mine workers' safety (i.e., it can create new hazards or worsen the existing ones) [3,4].

PE technology has various types of unique failure modes that are quite different from mechanical or hard-wired electronic systems traditionally employed in the mining area. Needless to say, the mining industry's experience with the functional safety of PE is quite different in comparison to other industrial sectors. This simply means that the application of the PE technology in the mining area, in regard to safety, must be considered with utmost care.

This chapter presents various important aspects of PE mining system safety.

## 12.2 PROGRAMMABLE ELECTRONICS USAGE TRENDS IN MINING INDUSTRY

Nowadays, the mining industry sector is increasingly using new technologies including programmable electronics due to various factors. An informal industry surveys' study, industry studies, and published equipment surveys revealed that the usage of PE in the area of mining is not limited to specific systems, mining methods, or commodities [5]. In fact, their usage can be classified under three basic areas, as shown in Figure 12.1 [5].

These basic areas are control, monitoring, and protection. Within the framework of these three basic areas, there are many application areas including [5]:

- Mine elevators and hoists.
- Mine atmospheric monitoring systems that monitor methane, fresh airflow, and carbon monoxide.
- Longwall coal mining systems.

DOI: 10.1201/9781003212928-12

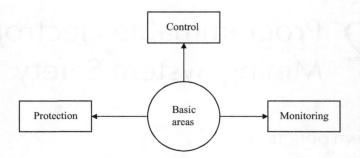

**FIGURE 12.1**   Basic areas of programmable electronics usage.

- Automated haulage vehicles for surface and underground metal/non-metal mines.
- Remote controllers for underground mining machines.

Underground mine atmospheric monitoring and control started in the late 1970s, and in the 1990s around 17% of all underground mines were equipped with computer-based systems [5]. Similarly, from 1990 to 1996, programmable longwall systems' usage doubled, i.e., to approximately 95% of all longwalls in the United States. Furthermore, microprocessor technology is increasingly being utilized in the monitoring and control of conveyor systems.

Nowadays, industry-wide trends are significantly inclined towards more use and complexity as machinery in general moves from localized PE control to distributed control of processes and machines. All in all, this trend is expected to increase in the future due to factors such as lower grades of coal and ores, economic-related pressures, and increasing degrees of difficulties in physically accessing such resources.

## 12.3   PROGRAMMABLE ELECTRONIC-ASSOCIATED MISHAPS

Nowadays, programmable-electronic-associated mishaps are an important issue in the mining industrial sector. The Mine Safety and Health Administration's (MSHA) serious thinking about the PE-based mining systems' functional safety began in 1990 as a result of the occurrence of an unplanned longwall shield mishap [6]. During the period 1995–2001, in the United States, a total of 11 PE-associated mining incidents were reported; four of these were fatalities [1,7]. During the same time period (i.e., 1995–2001), in New South Wales (NSW), Australia, 71 incidents were reported in underground coal mines [1].

A subsequent study of both sets of data, i.e., from the United States and Australia, clearly revealed that the majority of mishaps involved unexpected movements or startups of PE-based mining systems. In 1991, MSHA conducted a study of all long-wall installations in regard to PE. The study revealed that about 35% had experienced unexpected movements basically due to the following four problems [1,6]:

- **Problem I**: Software programming errors.
- **Problem II**: Water ingress.
- **Problem III**: Sticking or defective solenoid valves.
- **Problem IV**: Operator error.

A detailed analysis of the above United States and Australian data sets clearly reported the four major factors shown in Figure 12.2, which, directly or indirectly, contributed to PE-based mishaps [6,8]. However, it is to be noted that the solenoid valve-related problems were the leading factor contributing to PE-based mishaps.

In response to the above longwall-related mishaps, MSHA recommended improvements in items as follows:

- Maintaining alertness for abnormal operational sequences that might be indicative of a software-related problem.
- Operator training.
- Maintaining integrity of enclosure sealing.
- Timely maintenance.

Subsequently, MSHA realized the shortcomings of the above approach for all the complex PE mining systems and then proposed a safety framework which was largely based on the IEC 61508 safety life cycle [9].

## 12.4   LESSONS LEARNED IN ADDRESSING PROGRAMMABLE ELECTRONIC MINING SYSTEMS SAFETY

Over the years, the National Institute for Occupational Safety and Health (NIOSH), in conjunction with the Mine Safety and Health Administration and the industrial

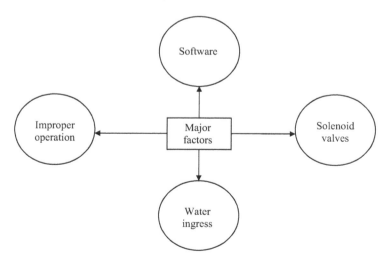

**FIGURE 12.2**   Major factors contributing to programmable-electronics-based mishaps.

sector, generated the NIOSH safety framework of PE mining systems' functional safety. In this regard, a number of valuable lessons were learned for properly addressing functional safety as well as for changing the perspectives and practices of the industrial sector [1]. These lessons are benefiting the mining industrial sector and can also be applied to other industrial sectors as well. Most of these lessons are as follows [1]:

- Highlight and understand all associated issues and perceptions.
- Involve the industrial sector as early as possible and on a continuous basis.
- Establish all important concepts, terminology, and definitions as early as possible.
- Carry out industry workshops as considered appropriate.
- Decompose the problem under consideration into effectively manageable parts.
- Make use of scenarios for conveying some types of information.
- Separate all the associated concerns.

## 12.5  METHODS FOR PERFORMING PROGRAMMABLE ELECTRONIC MINING SYSTEMS HAZARD AND RISK ANALYSIS

There are many methods that can be used to perform hazard and risk analysis of programmable electronic mining systems. The most useful ones are as follows [10]:

1. Interface analysis.
2. Hazard and operability studies (HAZOP).
3. Operating and support analysis (OASA).
4. Potential or predictive human error analysis.
5. Event tree analysis (ETA).
6. Action error analysis (AEA).
7. Sequentially timed events plot (STEP) investigation system.
8. Failure mode and effect analysis (FMEA).
9. Fault tree analysis (FTA).
10. Preliminary hazard analysis (PHA).

The first seven of the above ten methods are presented below, separately, and the remaining three methods are described in Chapter 4.

### 12.5.1  Interface Analysis

Interface analysis is used for determining the incompatibilities between subsystems and assemblies of an item/product that could, directly or indirectly, result in serious accidents. The analysis establishes that distinct parts/units can be integrated into a fairly viable system and a part's/unit's normal operation will not impair the performance or damage another part/unit or the whole system.

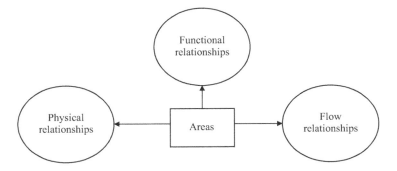

**FIGURE 12.3** Areas of relationships considered in interface analysis.

The relationships considered by the analysis can be grouped under three areas, as shown in Figure 12.3 [11,12].

Physical relationships are concerned with the items' physical aspects. For example, two items/units may be well designed and manufactured and function quite effectively individually, but they may fail to fit together due to dimensional differences or they may present other difficulties that may result in safety-associated problems. Functional relationships are concerned with multiple items. For example, in a situation where an individual item's outputs constitute the inputs to a downstream item, any deficiency in inputs and outputs may cause damage to the downstream item and, in turn, a safety hazard. The outputs could be in conditions such as unprogrammed, excessive, degraded, erratic, and zero outputs.

Flow relationships are concerned with two or more units where flow between these units may involve electrical energy, air, lubricant oil, fuel, steam, or water. The frequent flow-associated problems associated with many products are the flow's proper level of fluids and energy from one unit to another through confined passages. In turn, these may result, directly or indirectly, in safety-related problems.

Some of the benefits of the interface analysis are applicable to all types of systems and interfaces as well as at the subsystem to the component level. In contrast, some of its drawbacks are that it is difficult for applying to complex systems and it is quite difficult to discover all interface-associated incompatibilities for each and every operation.

### 12.5.2 HAZARD AND OPERABILITY STUDIES (HAZOP)

This is a systematic and structured qualitative method that had its beginnings in the chemical process industrial sector [4,10,12,13]. HAZOP has proven to be a very effective approach for highlighting unforeseen hazards designed into facilities for various reasons, or hazards introduced into already existing facilities due to factors such as changes made to process conditions or operating procedures.

Three fundamental objectives of HAZOP are as follows [12]:

- **Objective I**: Produce a complete process/facility description.
- **Objective II**: Review all facility/process elements for determining how deviations from the design intentions can occur.
- **Objective III**: Decide whether the above deviations can result in operating hazards/problems.

A HAZOP study can be conducted by following the five steps presented below [12,14]:

- **Step I: Establish study objectives and scope**. This is concerned with developing study objectives and scope by appropriately considering all relevant factors.
- **Step II: Form HAZOP team**. This involves forming a HAZOP team composed of individuals from design and operation with appropriate experience for determining all possible effects of deviations from intended applications.
- **Step III: Collect all types of relevant information**. This involves obtaining all the necessary documentation, drawings, and process description. More clearly, it includes items such as equipment-related specifications, layout drawings, process control-related logic diagrams, operating and maintenance procedures, and emergency response procedures.
- **Step IV: Analyze all major pieces of equipment and supporting items**. This involves conducting analysis of all major items of equipment as well as all supporting equipment, piping, and instrumentation using the documents obtained in step III.
- **Step V: Document the study**. This is the last step and is concerned with documenting items such as consequences of any deviation from the norm, a summary of deviations from the norm, and deviations' summary considered hazardous and credible.

Three main benefits of HAZOP are as follows [4]:

- Can produce comprehensive and detailed results.
- Very good track record of previous success and use.
- Does not require extensive training or specialized tools.

In contrast, two main drawbacks of HAZOP are it can be very time consuming for complex and large systems and it is best for short periods only because involved team members can lose effectiveness. HAZOP's application to a mining system is demonstrated in Ref. [4].

### 12.5.3 Operating and Supporting Analysis (OASA)

This method seeks to highlight potential hazards during operation and maintenance, find the related root causes, determine the acceptable risk level, and recommend necessary measures for risk reduction. The operating and support analysis is conducted by a group of persons familiar with the system's operation and interaction with all involved personnel.

Some of the factors considered during OASA are as follows [4,15]:

- Providing necessary documentation for the systems under consideration.
- Training all involved operation and maintenance personnel.
- Operation in normal and abnormal situations.
- Making all required changes to the system.
- Maintaining the equipment and its associated software.
- Testing of systems.

The main benefit of this method (i.e., OASA) is that it provides the identification of hazard in the context of the entire system operation. Its major drawback is that it requires a high degree of expertise concerning the system in question.

### 12.5.4  POTENTIAL OR PREDICTIVE HUMAN ERROR ANALYSIS

This is a team-based method and is quite similar to the hazard and operability studies (HAZOP) approach. However, it (i.e., method) focuses on human tasks and their associated error potential [16]. The method groups the human error causes into five basic classifications (along with their corresponding effects in parentheses): environment (adverse environments increase the likelihood of error), training (better training decreases the likelihood of error), stress (increases the likelihood of error), complexity (increases the likelihood of error), and fatigue (increases the likelihood of error).

Therefore, it is very important that the members of the team conducting the analysis consider all these causes of human error during the analysis. The basic analysis procedure is composed of the following two steps [4]:

- **Step 1**: Identify all key human tasks.
- **Step 2**: For each and every task, apply guide words such as incorrect selection, action applied to wrong interface object, incomplete action, incorrect action sequence, wrong action, incorrect action timing, and action omitted.

The main advantage of the method is that it can highlight a high proportion of potential errors. In contrast, its two main drawbacks are that it can be a quite time-consuming approach if there are many tasks and actions and its effectiveness very much depends on the team's expertise and effort [4].

### 12.5.5  EVENT TREE ANALYSIS (ETA)

This is a "bottom-up" approach for highlighting the possible outcomes when the occurrence probability of the initiating event is known. The method has proven to be an effective tool for analyzing facilities that have engineered accident-mitigating characteristics for highlighting the sequence of events that follow the initiating event and produce given sequences. Generally, it is assumed that each sequence event is either a success or a failure.

Because of the inductive nature of this approach, the fundamental question addressed is "What happens if … ?" Generally, ETA is used for performing analysis of more complex systems than those handled by the failure mode and effect analysis approach.

Two main benefits of ETA are that it is a quite effective tool for single events with multiple outcomes and for high risks not amendable to simpler analysis methods. In contrast, its drawbacks include an extremely time-consuming approach and that the event occurrence probabilities may be quite difficult to estimate.

Additional information on this method is available in Refs. [12,14,17].

### 12.5.6 ACTION ERROR ANALYSIS (AEA)

This is a quite effective method for identifying operator errors and their consequences. The method specifically focuses on the interactions between humans and a system during testing, operation, and maintenance phases. For operation and maintenance tasks, AEA is conducted by following the three basic steps shown in Figure 12.4 [15].

In step 3, for each and every action the following types of errors are considered:

- Actions applied to the wrong object.
- Incorrect actions taken.
- Temporal errors (i.e., action executed early or late).
- Wrong sequence of actions.
- Omission error (i.e., failure to an action).

The main benefit of AEA is that it is well suited for semi-automated or automated processes with operator interfaces; the main drawback is that it needs a high level of expertise concerning the system in question.

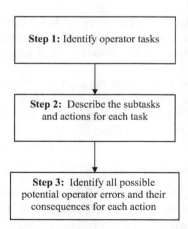

**FIGURE 12.4** Steps for conducting action error analysis for operation and maintenance tasks.

### 12.5.7 SEQUENTIALLY TIMED EVENTS PLOT (STEP) INVESTIGATION SYSTEM

STEP is a very useful analytical method that graphically shows sequentially timed events. All involved events are defined with formatted "building blocks" made up of an "actor and action" [4]. STEP analysis is very helpful for discovering and analyzing problems as well as assessing mitigation options. It is also used for performing analysis of the types and sequences of events that can result in an incident.

The two main benefits of STEP are that it can be used for defining and systematically analyzing complex processes or systems and that it facilitates focus-group analysis. In contrast, its main drawback is that it is generally perceived as a complicated approach that is costly to implement.

## 12.6 SOURCES FOR OBTAINING PROGRAMMABLE ELECTRONIC MINING SYSTEM SAFETY-RELATED INFORMATION

There are many sources for obtaining, directly or indirectly, programmable electronic mining system safety-associated information. This section presents a number of sources grouped under three classifications.

### 12.6.1 ORGANIZATIONS AND SYSTEMS

- Occupational Safety and Health Administration (OSHA), US Department of Labor, 200 Constitution Avenue, Washington, DC, USA.
- System Safety Society, 14252 Culver Drive, Suite A-261, Irvine, CA, USA.
- National Institute for Occupational Safety and Health (NIOSH), 200 Independence Avenue SW, Washington, DC, USA.
- National Safety Council, 444 North Michigan Avenue, Chicago, IL, USA.
- Institute of Electrical and Electronics Engineers (IEEE), 445 Hoes Lane, Piscataway, NJ 08854-4141, USA.
- The American Society of Safety Engineers, 1800 E. Oakton Street, Des Plaines, IL, USA.
- International Electro-technical Commission (IEC), 3 rue de Varembe, P.O. Box 131, CH-11, Geneva, Switzerland.
- GIDEP Data, Government Industry Data Exchange Program (GIDEP) Operations Center, Fleet Missile Systems, Analysis, and Evaluation, Department of Navy, Corona, CA, USA.
- National Electronic Injury Surveillance System, US Consumer Product Safety Commission, 5401 Westbard Street, Washington, DC, USA.
- Computer Accident/Incident Report System, System Safety Development Center, EG8G, PO BOX 1625, Idaho Falls, ID, USA.

### 12.6.2 COMMERCIAL SOURCES FOR OBTAINING STANDARDS

- Document Center, Inc., 111 Industrial Road, Suite 9, Belmont, CA 94002, USA.

- Global Engineering Documents, 15 Inverness Way East, Englewood, CO 80112-5704, USA.
- Total Information Inc., 844 Dewey Avenue, Rochester, NY 14613, USA.

### 12.6.3 STANDARDS AND BOOKS

- Defence Standard 00-58, Parts I and II, HAZOP Studies on Systems Containing Programmable Electronics, Directorate of Standardization, U.K. Ministry of Defence, Glasgow, UK, 1998.
- IEC 61508 SET, Functional Safety of Electrical/Electronic/Programmable Electronic Safety-Related Systems, Parts 1–7, International Electrotechnical Commission, Geneva, Switzerland, 2000.
- MIL-STD-882C, System Safety Program Requirements, US Department of Defense, Washington, DC, 1993.
- IEEE-STD-730, Standard for Software Quality Assurance Plans, Institute of Electrical and Electronics Engineers (IEEE), Piscataway, NJ, 1995.
- IEEE-STD-830, Recommended Practice for Software Requirement Specifications, Institute of Electrical and Electronics Engineers (IEEE), Piscataway, NJ, 1993.
- IEEE-STD-1228, Standard for Software Safety Plans, Institute of Electrical and Electronics Engineers (IEEE), Piscataway, NJ, 1994.
- Perrow, C., Normal Accidents: Living with High-Risk Technologies, Princeton University Press, Princeton, NJ, 1999.
- Levenson, N.G., Safeware: System Safety and Computers, Addison-Wesley, Reading, MA, 1995.
- Redmill, F., Chudleigh, M., Catmur, J., System Safety: HAZOP and Software HAZOP, Wiley, New York, 1999.
- Beizer, B., Software Testing Techniques, International Thomson Computer Press, London, 1990.

## 12.7  PROBLEMS

1. What are the basic areas of programmable electronics usage in mining?
2. What are the major factors contributing to programmable-electronics-based mishaps in the mining area?
3. Discuss the lessons learned in addressing programmable electronic mining systems safety.
4. List at least ten methods considered most useful for performing hazard and risk analysis of programmable electronic mining systems.
5. Describe the following two methods:
   - Event tree analysis.
   - Action error analysis.
6. What are the benefits and drawbacks of hazard and operability studies?
7. List the four most important sources for obtaining programmable electronic mining system safety-related information.
8. Describe potential or predictive human error analysis.

9. Compare event tree analysis with operating and support analysis.
10. What is a programmable electronic mining system?

## REFERENCES

1. Sammarco, J.J., Addressing the Safety of Programmable Electronic Mining Systems: Lessons Learned, Proceedings of the 37th IEEE Industry Applications Society Meeting, 2003, pp. 692–698.
2. Dhillon, B.S., *Mining Equipment Reliability, Maintainability, and Safety*, Springer-Verlag, London, 2008.
3. Sammarco, J.J., Kohler, J.L., Novak, T., Morley, L.A., Safety Issues and the Use of Software-Controlled Equipment in the Mining Industry, Proceedings of the 32nd IEEE Industry Applications Society Meeting, 1997, pp. 496–502.
4. Sammarco, J.J., *Programmable Electronic Mining Systems: Best Practice Recommendations (in nine parts)*, Report No. IC 9480 (Part 6: 5.1 System Safety Guidance), National Institute for Occupational Safety and Health (NIOSH), US Department of Health and Human Services, Washington, D.C., 2005. Available from NIOSH: Publications Dissemination, 4676 Columbia Parkway, Cincinnati, OH 45226, USA.
5. Sammarco, J.J., Safety Framework for Programmable Electronics in Mining, *Mining Engineering* 51, Vol. 12, 1999, pp. 30–33.
6. Dransite, G.D., Ghosting of Electro-hydraulic Long Wall Shield Advanced Systems, Proceedings of the 11th West Virginia University International Electrotechnology Conference, 1992, pp. 77–78.
7. *Fatal Alert Bulletins, Fatal Grams and Fatal Investigation Reports*, Mine Safety and Health Administration (MSHA), Washington D.C., May 2001, retrieved from website: www.msha.gov/fatals/fab.htm.
8. Wandby, J.F., *Underground Coal Mining Remote Control of Mining Equipment: Known Incidents of Unplanned Operation in New South Wales (NSW) Underground Coal Mines*, Department of Mineral Resources, NSW Department of Primary Industries, Maitland, NSW, Australia, 2001.
9. IEC61508, Parts 1–7, Functional Safety of Electrical/Electronic/Programmable Electronic Safety-related Systems, International Electrotechnical Commission, Geneva, Switzerland, 1998.
10. *Defense Standard 00–58 (Parts I and II), HAZOP Studies on Systems Containing Programmable Electronics*, Directorate of Standardization, UK Ministry of Defense, Glasgow, UK, 1998.
11. Hammer, W., *Product Safety Management and Engineering*, Prentice-Hall, Englewood Cliffs, NJ, 1980.
12. Dhillon, B.S., *Engineering Safety: Fundamentals, Techniques, and Applications*, World Scientific, River Edge, NJ, 2003.
13. Redmilll, F., Chudleigh, M., Catmur, J., *System Safety: HAZOP and Software HAZOP*, Wiley, New York, 1999.
14. CAN/CSA-Q6340–91, *Risk Analysis Requirements and Guidelines*, prepared by the Canadian Standards Association, Toronto, 1991. Available from the Canadian Standards Association (CSA), 178 Rexdale Boulevard, Rexdale, Ontario, Canada.
15. Harm-Ringdahl, L., *Safety Analysis: Principle and Practice*, Elsevier, London, 1993.
16. American Institute of Chemical Engineers, *Guidelines for Preventing Human Error in Process Safety*, Center for Chemical Process Safety, American Institute of Chemical Engineers, New York, 1994.
17. Cox, S.J., Tait, N.R.S., *Reliability, Safety, and Risk Management*, Butterworth-Heinemann, Oxford, 1991.

# 13 Safety in Offshore Oil and Gas Industry

## 13.1 INTRODUCTION

The history of the offshore oil and gas industry goes back to around 1891, when the first submerged oil wells were drilled from platforms on piles in the Grand Lake St. Marys freshwaters in Ohio, United States [1]. Over the past six decades, offshore production has increased tremendously, currently around 30% of the entire world oil and gas production coming from offshore [2,3].

Offshore industrial sector has become an important element of the industry as each year a vast sum of money is spent on offshore-associated developments around the globe. In order to meet the significantly increasing demand for oil and gas, the industrial sector uses and develops leading-edge technology to drill even deeper.

Over the years, many accidents in the offshore industrial sector have occurred and caused many fatalities and a large amount of money being spent on damages. Some examples of the deadliest accidents in the offshore oil and gas industrial sector are the Alexander L. Kielland accident in Norway in 1980, the Piper Alpha platform accident in the United Kingdom in 1998, and the Mumbai High North Platform accident in India in 2005 [4].

Needless to say, safety has become a very important issue in the offshore oil and gas industrial sector. This chapter presents various important aspects of safety in offshore oil and gas industry.

## 13.2 OFFSHORE INDUSTRIAL SECTOR-RELATED RISK PICTURE

The offshore industrial sector "risk picture" is a highly multifaceted one. Therefore, establishing an actual "factual risk picture" is a quite difficult task due to factors such as follows [5]:

- A basic uncertainty in extrapolating information from the past to present risk.
- Inconsistency in recording and reporting incidents across the industrial sector.
- Changes in the regulatory regime and in the industrial sector as well as its management systems.
- Different measures of risk.

Nonetheless, the main contributors to individual risk are shown in Figure 13.1 [5].

DOI: 10.1201/9781003212928-13

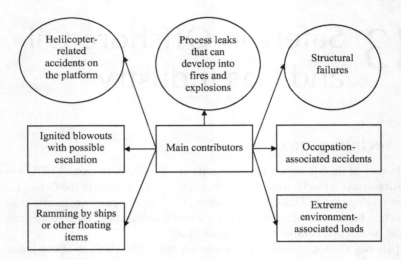

**FIGURE 13.1**   Main contributors to offshore individual risk.

Additional information on offshore industrial sector risk picture is available in Ref. [5].

## 13.3   OFFSHORE WORKER SITUATION AWARENESS CONCEPT, STUDIES, AND THEIR RESULTS

Past experiences over the years clearly indicate that there is a definite need for maintaining the situation awareness of workers at a high level for properly insuring their operations' safety in many industrial areas. Situation awareness may be expressed as the perception of the elements in the environment within the framework of a volume of time and space, the comprehension of their proper meanings, and the projection of their status in the near future [6]. The three main elements of situation awareness are as follows [6]:

- **Element I: Factors affecting situation awareness**: Two factors that, directly or indirectly, affect situation awareness are workload and stress. Unusually high or low workloads are considered to potentially impact humans' performance to a certain degree [7]. Low workload can result in boredom with consequent inattentiveness, lower vigilance, and significantly reduced motivation. Furthermore, when less attention is being given to workplace-related situations or conditions that, in turn, can result in poor situation awareness.

    On the other hand, high workloads can result in impairing situation awareness of workers as they may not be completely aware of situation-associated changes, and can make incorrect decisions on the basis of wrong or incomplete information. Furthermore, there is some evidence that increments in workload have quite detrimental impacts to a certain degree on offshore workers' psychological well-being [6–8].

- **Element II: Situation awareness levels**: These are concerned with perception, comprehension, and projection. Perception calls for monitoring the surrounding environment on a continuous basis for encoding sensory-associated information as well as detecting changes in significant stimuli.

  Comprehension involves combination, interpretation, storage, and retention of incoming information for forming a picture of the current condition or situation whereby the events'/objectives' significance is understood. Finally, projection is the result of comprehension and perception, and it is concerned with predicting all possible future events/states.

- **Element III: Team situation awareness**: This is concerned with teamwork, as the successful accomplishment of a stated task (e.g., a drilling task in the offshore oil and gas industry) is totally dependent upon all the crew members collectively working together. Therefore, it is absolutely essential for all involved crew members to have a clear mutual comprehension of the situation under consideration.

  More clearly and in short, it may simply be stated that all involved crew members must have a situation awareness (this shared awareness is known as team situation awareness) [4,6].

## 13.3.1 OFFSHORE WORKER SITUATION AWARENESS-ASSOCIATED STUDIES AND THEIR RESULTS

Over the years, many studies concerning offshore workers' situation awareness have been conducted. Two such studies and their results are presented below, separately.

### 13.3.1.1 Study I

This study was concerned with situation awareness errors in offshore drilling-related incidents. The study reported three classifications of such errors along with their occurrence percentages as shown in Figure 13.2 [6].

The breakdown of the perception-associated incident errors along with occurrence percentages was as follows [4,5]:

- Memory loss: 0.1%.
- Data not available: 9.7%.
- Misconception of data: 14.2%.
- Hard to discriminate or detect data: 15.7%.
- Failure to monitor or observe data: 26.8%.

Similarly, the breakdown of comprehension-associated incident errors along with their occurrence percentages was as follows [4,6]:

- Overreliance on default values: 2.2%.
- Lack of/poor mental model: 6.7%.
- Use of incorrect mental model: 11.1%.

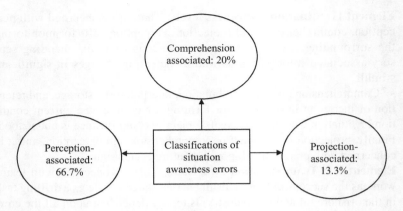

**FIGURE 13.2** Classifications of situation awareness errors in offshore drilling incidents and their corresponding occurrence percentages.

Finally, the breakdown of projection-associated incident errors along with occurrence percentages was as follows [4,6]:

- Over projection of current trends: 0.1%.
- Lack of/poor mental model: 13.2%.

### 13.3.1.2  Study II

This study consists of interviews with all personnel involved with offshore drilling, and the aim of the study was to comprehend how well the situation awareness concept is recognized within the offshore industrial sector. During the interviews, the following questions were asked [6]:

- What can be done to check the awareness of workers?
- How is situation awareness known in the offshore industry?
- How can reduced awareness be improved?
- What factors affect the quality of a person's awareness?
- How is team situation awareness achieved?
- What are the indicators of reduced awareness?

The question "What can be done to check the awareness of workers?" received four responses. Two of these responses were constant assessment of surroundings and risk. The question "How is situation awareness known in the offshore industry?" received three responses. They were safety awareness, safety accountability, and positional awareness.

The question "How can reduced awareness be improved?" received ten responses. They were discussion of events, training, increased involvement in rig activities, alter the crew line-up, alter the work level, placing them (personnel) in a different job, interaction, problem-solving, communication, and removal from the situation.

The question "What factors affect the quality of a person's awareness?" received 12 responses. These responses were as follows [4,6]:

1. Stress and workload.
2. Weather/seasons.
3. Communication (good and bad).
4. Fatigue.
5. Having a near miss.
6. Experience (and new personnel).
7. Routine task/complacency.
8. Supervisory responsibility.
9. Home/family problems.
10. Job prospects.
11. Daydreaming.
12. Conflict.

The question "How is team situation awareness achieved?" received eight responses. These responses were as follows [4,6]:

1. Experience
2. Understand capabilities and traits planning
3. Cooperation
4. Adaptability
5. Trust
6. Time
7. Increased interaction
8. Consistency

Finally, the question "What are the indicators of reduced awareness?" received five responses. These responses were as follows [4,6]:

1. Reduction in communication
2. Repetition of instructions
3. Reduced work standards
4. Character change
5. An expressionless appearance

Additional information on responses to all the above six questions is available in Ref. [6].

## 13.4 OFFSHORE INDUSTRY DEADLIEST ACCIDENTS' CASE STUDIES

Over the years, a large number of accidents in the offshore industrial sector have occurred around the globe. Eight of the deadliest of these accidents are described below, separately.

### 13.4.1 Piper Alpha Accident

This accident is considered the deadliest offshore oil rig accident in history. Piper Alpha was a North Sea oil production platform, and it was located around 120 miles northeast of Aberdeen, United Kingdom. The platform was operated by Occidental Petroleum (Caledonia) Ltd and it became operational in 1976. Initially, the platform was constructed for producing crude oil, but later on, with the installation of gas conversion equipment, it also started producing gas.

Piper Alpha produced oil and gas from its 24 wells for delivery to the Flotta oil terminal located on the Orkney Islands as well as to other installations through three separate pipelines. At the time of disaster's occurrence, the platform produced approximately 10% of North Sea oil and gas [9–12]. On July 6, 1988, due to gas leakage from one of the condensate pipes at the platform, explosions and a resulting fire destroyed the platform and caused 167 casualties [10,11].

A subsequent investigation into the disaster was carried out by the United Kingdom government that highlighted a number of factors that, directly or indirectly, contributed to the Piper Alpha incident's severity. Two of these factors were as follows [10–12]:

- Lack of blast walls and existence of firewalls. More clearly, the existing firewalls predated the gas conversion equipment's installation as well as were not properly upgraded to blast walls subsequent to the installation.
- Serious breakdown in the chain of command as well as lack of any proper communication to the platform's all crew members.

The investigation made 106 recommendations for changes to the existing North Sea safety-associated procedures. The offshore industrial sector accepted all the recommendations. Additional information on the Piper Alpha accident is available in Refs. [9–13].

### 13.4.2 Alexander L. Kielland Accident

Alexander L. Kielland was a Norwegian semisubmersible rig/platform in the Ekofisk oil field, Norwegian continental shelf, approximately 235 miles east of Dundee, Scotland, United Kingdom. The rig/platform was owned by the Stavanger Drilling Company of Norway, and it was named after a Norwegian writer. At the time of the disaster's occurrence, the rig/platform was hired by a US company called Phillips Petroleum.

After about 40 months of service, the rig/platform was no longer utilized for drilling, but it served as a so-called flotel (i.e., a floating hotel) for workers from the close-by Edda platform. On March 27, 1980, wind gusts of around 40 knots created waves up to 12 m high that, in turn, caused the rig/platform to collapse into the North Sea, and caused 123 deaths of off-duty workers.

A subsequent investigation carried out by the Norwegian government concluded that the rig collapsed due to a fatigue crack in one of the rig's six bracings (bracing D-6), which connected the collapsed D-leg to the rest of the rig [4,14]. Additional information on this accident is available in Refs. [4,13,14].

### 13.4.3  GLOMAR JAVA SEA DRILLSHIP ACCIDENT

The Glomar Java Sea Drillship was built in 1975 for drilling wells down to around 25,000 ft in water depths of approximately up to 1,000 ft. The drillship was designed by Global Marine, Inc., and constructed by the Livingston Shipbuilding company of Orange, Texas. The 40-ft-long drillship was contracted to ARCO China, and it arrived in the South China Sea in January 1983 [13].

On October 25, 1983, the drillship capsized and sank in the South China Sea, at the depth of 317 ft, around 63 nautical miles south-west of Hainan Island, China and approximately 80 nautical miles east of the Socialist Republic of Vietnam. The incident caused the death of 81 persons on board the drillship.

A subsequent investigation carried out by the United States National Transportation Safety Board concluded that the most likely cause for the Glomar Java Sea Drillship's sinking and capsizing was the flooding of its starboard tanks 6 and 7 through a hull fracture during typhoon Lex [15]. Additional information on this accident is available in Refs. [13,15].

### 13.4.4  SEACREST DRILLSHIP ACCIDENT

The Seacrest Drillship accident occurred on November 3, 1989, in the South China Sea approximately 430 km south of Bangkok, Thailand. At the time of the accident, the drillship was anchored for drilling at the Platong gas field managed and owned by Unocal [13]. On the accident day, Typhoon Gay produced approximately 40 ft high waves that capsized the drillship.

The accident caused the death of 91 crew members. Although on November 4, 1989, the drillship was reported missing, it was only discovered floating upside down the following day by a search helicopter. It is believed that the drillship's capsize took place so fast that there was no distress signal and no time left for the crew members to respond to the accident.

Additional information on this accident is available in Ref. [13].

### 13.4.5  BOHAI 2 OIL RIG ACCIDENT

The Bohai No. 2 oil rig was located in the Gulf of Bohai off the coast of China, and it was operated and managed by the Ocean Oil Company, China Petroleum Department. The rig was a self-evaluating drilling unit that sank on November 25, 1979, while being towed after encountering a storm with force 10 winds [3,4]. The incident resulted in the deaths of 72 out of 76 persons onboard the rig.

The post-disaster investigations into the accident attributed many causes for its severity. Four of these causes are as follows [4,16]:

- **Cause I**: Failure to properly stow deck equipment prior to towing.
- **Cause II**: Failure to properly follow standard tow-associated procedures in regard to weather.
- **Cause III**: Poor training of crew members in regard to the use of lifesaving-associated equipment.
- **Case IV**: Poor emergency evacuation-associated procedures.

Additional information on this accident is available in Refs. [13,16–18].

### 13.4.6   ENCHOVA CENTRAL PLATFORM ACCIDENT

The Enchova Central Platform was located in the Campos Basin close to Rio de Janeiro, Brazil, and it was operated by the Brazilian company called Petrobras. The accident occurred on August 16, 1984, due to a blowout, which, in turn, caused an explosion and fire at the central platform. Although most of the workers from the platform, were evacuated safely from the platform by helicopter and lifeboats, 42 of them lost their lives during the evacuation process [4,13].

More clearly, in this case, the most serious incident took place when the lowering mechanism of a lifeboat malfunctioned. Consequently, the lifeboat remained vertically suspended until the stern support brake and the lifeboat fell around 20 m deep into the sea and killed 36 of its occupants. Another six persons got killed as they jumped around 40 m from the platform into the sea.

Additional information on this accident is available in Refs. [13,19].

### 13.4.7   MUMBAI HIGH NORTH PLATFORM ACCIDENT

The Mumbai high field, found in 1974, is the largest oil and gas field in India. The field is located in the Arabian Sea, around 100 miles west of Mumbai coast. The Mumbai High North Platform was constructed in 1981 and was an oil and natural gas processing complex operated and owned by India's state-owned Oil and Natural Gas Corporation [13].

The complex produced approximately 120,000 barrels of oil and around 4.4 million cubic meters of gas per day [13]. The platform was a seven-storey high steel structure, and it contained five gas export risers as well as ten fluid import risers that were located just outside its jacket. On July 25, 2005, a multipurpose support vessel (MSV) called "Samudra Surakasha" collided with the platform that caused rupture of one or more of the platform's gas export risers. The reluctant gas leakage caused ignition that set the platform on fire, and the heat radiation resulted in damage to the Noble Charlie Yester jack-up rig engaged in drilling-related operations close to the platform. Furthermore, heat radiation also caused damage to MSV.

All in all, the platform was destroyed within hours and resulted in 22 fatalities. Additional information on this accident is available in Refs. [20,21].

### 13.4.8   BAKER DRILLING BARGE ACCIDENT

The Baker Drilling Barge accident took place on June 30, 1964, in the Gulf of Mexico. Fire and an explosion on the drilling barge caused 21 deaths and 22 injuries [13]. At the time of the occurrence of the accident, the Baker Drilling Barge was deployed for drilling operation for Pan American Petroleum Corporation in Eugene Island, Gulf of Mexico.

In the morning of June 30, 1964, the drilling barge's two 260-ft-long hulls suffered a severe blowout. Just in few minutes after the blowout, the entire drilling barge was engulfed with fire and explosion. After heeling the aft for around half-hour, the vessel sank upside down in the water.

Additional information on this accident is available in Ref. [13].

## 13.5 OFFSHORE INDUSTRY ACCIDENT REPORTING APPROACH AND OFFSHORE ACCIDENT-ASSOCIATED CAUSES

Generally, the accident reporting approach utilized in the offshore industrial sector can be divided into the following two areas [4,22]:

- **Area I: Serious accidents or incidents**: In this case, members of the investigation teams along with government-appointed accident inspectors fly from onshore to the offshore site. Generally, all accidents/incidents are processed and documented through stated channels and the accident reports' final copies are distributed to all the concerned bodies and authorities.
- **Area II: Minor accidents**: In this case, the investigators are safety officers and supervisors who generally have some training in the area. As the need arises, an investigation team is formed for conducting a more comprehensive investigation into the accident occurrence.

Generally the forms used for accident reporting contain information on items such as follows [4,22]:

- Location of occurrence, date, and time.
- Equipment failures.
- Contributory factors (e.g., environmental conditions, any existing hazards).
- Immediate and underlying causes.
- Type of incident/accident: injury, property damage, flammable or poisonous substance leaks, process disruption, material loss, explosion or fire, hazards, environmental harm, dangerous occurrences, disease, and near misses.
- All permits being issued and procedures being followed.
- Protective clothing being worn by all involved personnel.
- Equipment being used, including safety devices and equipment.
- Personal details of all personnel involved, including a supervisor at the time of the accident/incident occurrence.
- The type of work being performed and experience of all involved individuals.
- Other people performing their tasks in the surrounding area.

### 13.5.1 OFFSHORE ACCIDENT-ASSOCIATED CAUSES

A study of the accident reporting forms used by 25 companies in the United Kingdom reported that there were a large number of immediate causes for accidents' occurrence [22]. Most of these immediate causes were as follows [4,22]:

- Equipment used improperly.
- Wrong speed.
- Use of defective equipment.
- Adjusted equipment in operation.
- Failure to warn/secure.
- Operating without proper authority.
- Proper equipment not used.
- Workers under the influence of drugs/alcohol.
- Safety equipment/device made inoperable.
- Work carried out on live or unsafe equipment.
- Lack of attention/forgetfulness.
- Serviced equipment in operation.
- Improper loading/lifting.

In addition, there were also a large number of underlying causes for accidents' occurrence [22]. All these underlying causes are grouped under two classifications as shown in Figure 13.3 [4,22].

The personal factors classification has the following four elements [4,22]:

- **Element I: Knowledge and skill**: Its subelements were poor training, lack of education, lack of hands-on instructions, lack of experience, poor orientation, inadequate practice, misunderstood directions, lack of job instructions, and lack of awareness.
- **Element II: Improper motivation**: Its subelements were inappropriate attempt to save time, peer pressure, attitude, recklessness, lack of anticipation, inadequate thought and care, inattention, aggression, and horseplay.
- **Element III: Capability**: Its subelements were poor judgement, inability to comprehend, lack of competence, lack of physical capability, lack of mental capability, concentration demands, perception demands, judgement demands, and memory failure.
- **Element IV: Stress**: Its subelements were fatigue, frustration, health hazards, monotony, and general stress.

Similarly, the job factors classification has the following four elements [4, 22]:

**FIGURE 13.3**   Classifications of underlying causes for the occurrence of offshore accidents.

- **Element I: Task**: Its subelements were poor equipment selection, inappropriate matching of individual to job task, inadequate work planning, inadequate or no job description, confusing directions, conflicting goals, failure in communication, and time problems.
- **Element II: Supervision**: Its subelements were poor work planning, insufficient supervisory job knowledge, improper production incentives, lack of inspections, poor supervisory examples, unclear responsibilities, incomplete instruction and training, and inadequate discipline.
- **Element III: Organization**: Its subelements were company policy, poor staffing and resources, poor safety plan, working hour policies, poor procedures, safety systems, adequacies of systems, and competence standards.
- **Element IV: Management**: Its subelements were poor planning, bad examples set by management, management practices, management job knowledge, communication, and qualification and experience criteria.

Additional information on all the above elements is available in Refs. [22,23].

## 13.6   PROBLEMS

1. Discuss the industrial sector-related risk picture.
2. Describe the offshore worker situation awareness concept.
3. Describe the Piper Alpha accident.
4. What was the name of the company that constructed the Glomar Java Sea Drillship and how many deaths occurred in the following accidents?
   - Alexander L. Kielland accident.
   - Mumbai High North Platform accident.
   - Bohai 2 Oil Rig accident.
5. Describe the Baker Drilling Barge accident.
6. Discuss offshore accident-associated causes.
7. Describe offshore industry accident reporting approach.
8. Compare the Mumbai High North Platform accident with the Piper Alpha accident.
9. What are the items on which offshore accident reporting forms normally contain information?
10. Write an essay on safety in offshore oil and gas industry.

## REFERENCES

1. Offshore Drilling, retrieved on June 19, 2015 from website: https://en.wikipedia.org/wiki/offshore-drilling (last modified on January 13, 2016).
2. About Offshore Oil and Gas Industry, retrieved on June 19, 2015 from website: http://www.modec.com/about/industry/oil-gas.html.
3. Dhillon, B.S., *Safety and Reliability in the Oil and Gas Industry: A Practical Approach*, CRC Press, Boca Raton, Florida, 2016.
4. Dhillon, B.S., *Mine Safety: A Modern Approach*, Springer-Verlag, London, 2010.

5. Tveit, O.J., Safety Issues on Offshore Process Installation: An Overview, *Journal of Loss Prevention in the Process Industries*, Vol. 7, No. 4, 1994, pp. 267–272.
6. Sneddon, A., Mearns, K., Flin, R., Situation Awareness and Safety in Offshore Drill Crews, *Cognition, Technology, and Work*, Vol. 8, No. 4, 2006, pp. 255–267.
7. Parkes, K., Psychosocial Aspects of Stress, Health and Safety on North Sea Installations, *Scandinavian Journal of Work, Environment, and Health*, Vol. 24, No. 5, 1988, pp. 321–333.
8. Sutherland, K., Flin, R., Stress at Sea: A Review of Working Conditions in the Offshore Oil and Fishing Industries, *Work Stress*, Vol. 3, 1989, pp. 269–285.
9. Petrie, J.R., *Piper Alpha Technical Investigation Interim Report*, Petroleum Engineering Division, Department of Energy, London, U.K., 1988.
10. Pate-Cornell, M.E., Learning from the Piper Alpha Accident: Analysis of Technical and Organizational Factors, *Risk Analysis*, Vol. 13, No. 2, 1993, pp. 215–232.
11. Pate-Cornell, M.E., Risk Analysis and Risk Management for Offshore Platforms: Lessons from the Piper Alpha Accident, *Journal of Offshore Mechanics and Arctic Engineering*, Vol. 115, No. 1, 1993, pp. 179–190.
12. Hull, A.M., Alexander, D.A., Klein, S., Survivors of the Piper Alpha Oil Platform Disaster: Long-Term Follow up Study, *The British Journal of Psychiatry*, Vol. 181, 2002, pp. 433–438.
13. The World's Worst Offshore Oil Rig Disasters, retrieved on January 20, 2015 from website: http://www.offshore-technology.com/features/feature-the-worlds-deadliest-offshore-oil-rig.
14. Alexander L. *Kielland Accident, Report of a Norwegian Public Commission Appointed by Royal Decree of March 28, 1980*, Report No. ISBN B0000ED27N, Norwegian Ministry of Justice Police, Oslo, Norway, March 1981.
15. *Capsizing and Sinking of the United States Drillship Glomar Java Sea*, Report No. NTBS-MAR-84-08, National Transportation Safety Board, Washington, DC, 1984.
16. Bohai 2 Jack-Up, retrieved on April 10, 2009 from website: http://home.versa-tel/.nl/the-sims/rig/bohai 2.htm.
17. Santos, R.S., Feijo, L.P., Safety Challenges Associated with Deepwater Concepts Utilized in the Offshore Industry, Proceedings of the 9th International Symposium on Maritime Health, 2007, pp. 1–8.
18. Santos, R.S., Feijo, L.P., Deepwater Safety Challenges to Consider in a Fast-paced Development Environment, *Offshore*, Vol. 68, No. 3, 2008, pp. 1–5.
19. One Hundred Largest Losses, Marsh Risk Consulting, retrieved on April 10, 2009 from website: http://www.marshriskconsulting.com/st/PSEV-C-352-NR-304-htm.
20. *Riser Safety in UK Waters-Lessons from Mumbai High North Disaster*, Report No. SPC/Technical/OSD/33, Hazardous Installations Directorate, Health and Safety Executive, London, May 2006.
21. Mumbai High North, retrieved on April 10, 2009 from website: http://home.versatel.nl/the-sims/rig/mhn.htm.
22. Gordon, R.P.E., The Contribution of Human Factors to Accidents in the Offshore Oil Industry, *Reliability Engineering and System Safety*, Vol. 61, 1989, pp. 95–108.
23. Bird, F.E., Germain, G.L., *Practical Loss Control Leadership: The Conservation of People, Property, Process, and Profits*, Institute Publishing (ILCI), Longville, Georgia, 1989.

# 14 Software Safety

## 14.1 INTRODUCTION

Nowadays, computers have become an important element of day-to-day life and they are made up of both hardware and software components. Each year a vast sum of money is spent to develop various types of software around the globe, and safety has become a very important issue. More specifically, in many applications, proper functioning of software is so crucial that a simple malfunction may result in a large-scale loss of lives and a high cost. For example, commuter trains in Paris, France, serve approximately 800,000 passengers daily and depend on software signalling [1].

This chapter presents various important aspects of software safety.

## 14.2 SOFTWARE SAFETY-ASSOCIATED FACTS, FIGURES, AND EXAMPLES

Some of the facts, figures, and examples directly or indirectly concerned with software safety are as follows:

- The software industrial sector in the United States is worth at least US$300 billion per annum [2].
- Over 70% of the companies involved in the software development-related business develop their software by using ad hoc and unpredictable methods/techniques [3].
- During the Gulf War, a software error shut down a patriot missile system. Consequently, an enemy SCUD missile killed 27 persons and wounded 97 [4,5].
- A software error in a French meteorological satellite led to the destruction of 72 of the 141 weather balloons [6].
- An instrument failure due to a safety-associated software issue caused the SAAB JAS39 Gripen fighter plane to crash [4].
- Software errors in a computer-controlled therapeutic radiation machine known as Therac 25 caused deaths of two patients and severe injuries to another patient [7–10].
- Due to software errors two persons died in incidents involving heart pacemakers [8].
- A software error caused a radioactive heavy water spill at a Nuclear Power Generating Station in Canada [4].
- A software error resulted in the failure of a computer-aided dispatch system for the London Ambulance Service by sending an incorrect ambulance to an incident [11].

DOI: 10.1201/9781003212928-14

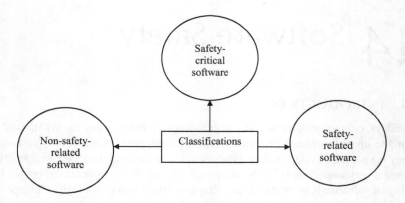

**FIGURE 14.1**   Software safety classifications.

## 14.3   SOFTWARE SAFETY CLASSIFICATIONS AND WAYS THAT SOFTWARE CAN CONTRIBUTE TO HAZARDS

Software safety may be categorized under three classifications as shown in Figure 14.1 [12]. These classifications are safety-critical software, safety-related software, and non-safety-related software. The safety-critical software controls or conducts such functions, and if they are executed erroneously or if they failed to execute appropriately, they could directly inflict serious injuries to humans and/or the environment and cause fatalities. Similarly, the safety-related software controls or conducts such functions which are activated for minimizing or preventing altogether the effect of a safety-critical system failure.

Finally, the non-safety-related software controls or conducts those system functions that are not concerned with safety. All in all, most mission-critical systems incorporate a combination of non-safety-related, safety-related, and safety-critical software [12].

There are many ways in which software can cause/contribute to a hazard. The main ones are as follows [6,13]:

- Failure to perform a necessary function.
- Poor response to a contingency.
- Conducting a required function out-of-sequence.
- Failure to recognize a hazardous condition requiring a corrective action.
- Providing incorrect solution to a problem.
- Poor timing of response for an adverse situation.
- Conducting an unnecessary function.

## 14.4   BASIC SOFTWARE SYSTEM SAFETY-ASSOCIATED TASKS

There are many software system safety-associated tasks. Some of the basic ones are as follows [14]:

- Develop a tracking system within the software framework along with system configuration control structure for assuring the traceability of safety requirements and their flow through documentation.
- Trace all types of safety-related requirements and constraints right up to the code.
- Highlight with care all the safety-critical elements and variables for use by code developers.
- Develop on the basis of highlighted software system safety constraints the system-specific software design criteria and requirements, computer–human interface associated requirements, and testing requirements.
- Review with care the test results concerning safety issues and trace the highlighted safety-associated software problems right back to the system level.
- Clearly show the software system safety-associated constraint consistency in regard to the software requirements specification.
- Highlight the components of the software that control safety-critical operations and then direct all necessary safety analysis and tests on those particular functions and on the safety-critical path that leads to their execution.
- Conduct any special safety-associated analyses, e.g., computer–human interface analysis or software fault tree analysis.
- Establish appropriate safety-associated software test plans, test case requirements, test descriptions, and test procedures.
- Trace all highlighted system hazards to the hardware-software interface.

## 14.5 SOFTWARE SAFETY ASSURANCE PROGRAM AND SOFTWARE QUALITY ASSURANCE ORGANIZATION'S ROLE WITH RESPECT TO SOFTWARE SAFETY

A software safety assurance program within the framework of an organizational setup basically involves the following three maturity levels [4,5]:

- **Maturity level I**: This is concerned with the development of an organization's/company's culture that clearly recognizes the importance of safety-associated issues. More specifically, company software developers conduct their tasks according to standard development rules and apply them quite consistently.
- **Maturity level II**: This is concerned with the implementation of a development process that involves safety assurance reviews as well as hazard analysis for highlighting and eliminating safety-critical situations prior to being designed into the system.
- **Maturity level III**: This is concerned with the utilization of a design process that documents results as well as implements continuous improvement methods for eliminating safety-critical errors in the system software.

Some of the items that need to be considered with care during a software safety assurance program's implementation are as follows [5]:

- Software system safety is addressed in regard to a team effort that involves groups such as management, quality assurance, and engineering.
- All software system-associated hazards are highlighted, evaluated, tracked, and eliminated as per requirements.
- Past software safety-related data are considered with care as well as used in all future software development projects.
- All human–computer interface requirements and software system safety-related requirements are consistent with contract requirements.
- Software system safety is clearly quantifiable to the stated level of risk using the general measuring methods.
- Software system safety-related requirements are specified and developed as an element of the organization's design policy.
- Changes in mission requirements, configuration, or design are conducted such that they clearly maintain an acceptable level of risk.

A software safety assurance program must also consider factors as follows [4,5]:

- Minimize the retrofit actions required for improving safety by including all appropriate safety features during research and development in an effective manner.
- Minimize risk as much as possible when using and accepting new designs, test and production methods, materials, etc.
- Ensure with care that safety is designed into the system cost effectively and in a timely fashion.
- Conduct necessary changes in design configuration and user-related requirements such that they maintain an acceptable risk level.
- Record all types of safety-associated data.

A software quality assurance organization plays various roles in regard to software safety. Some of these roles are presented below [4,5]:

- Establish the operational safety-related policy that clearly highlights acceptable risks and operational alternatives to hazardous operations.
- Define all appropriate requirements for performing operational safety reviews.
- Carry out safety audits and reviews of operational systems on a regular basis.
- Define user-safety-associated requirements, the operational doctrine, and the operational concept.
- Approve the findings of safety testing prior to releasing the systems.
- Evaluate, investigate, resolve, and document all reported safety-associated operational incidents.

- Chair operational safety-associated review panels.
- Determine appropriate safety-associated criteria for system acceptance.

## 14.6 USEFUL SOFTWARE SAFETY DESIGN-ASSOCIATED GUIDELINES

There are many useful software safety design-associated guidelines developed over the years by professionals working in the field of computers. A careful application of such guidelines can be very useful for improving software safety. Some of these guidelines are as follows [5,15]:

- Ensure that all conditional statements satisfy all possible conditions effectively and are under full software control.
- Include an operator for authorizing or validating the execution of safety-critical commands.
- Develop appropriate software modules for monitoring critical software in regard to faults, errors, timing problems, or hazardous states.
- Incorporate all necessary provisions to detect and log system errors.
- Develop software design such that it quite effectively prevents inadvertent/ unauthorized access and/or any modification to the code.
- Include appropriate mechanisms for ensuring that safety-critical computer software parts and interfaces are under positive control at all times.
- Include the requirement for a password along with confirmation prior to the execution of a safety-critical software module.
- Remove unnecessary or obsolete code.
- Separate and isolate all safety-critical software modules from non-safety-critical software modules.
- Avoid allowing safety-critical software patches throughout the development process.
- Avoid using all 1 or 0 digits for critical variables.
- Initialize spare memory with a bit pattern that, if ever accessed and executed, will clearly direct the involved software toward a safe state.

## 14.7 SOFTWARE HAZARD ANALYSIS METHODS

There are many methods that can be used for performing various types of software hazard analysis. Most of these methods are as follows [5, 16–19]:

- Event tree analysis.
- Software sneak circuit analysis.
- Software fault tree analysis.
- Proof of correctness.
- Failure modes and effect analysis.
- Code walk-through.
- Hazard and operability studies.

- Cause-consequence diagrams.
- Desk checking.
- Design walk-through.
- Monte Carlo simulation.
- Petri net analysis.

The first six of the above methods are described below [5, 16–19].

### 14.7.1 Event Tree Analysis (ETA)

This is a quite useful method and it models the sequence of events resulting from a single initiating event. In regard to its application to software, the initiating event is taken from a code segment considered safety critical (i.e., suspected of error or code inefficiencies). Generally, ETA assumes that each sequence event is either a success or a failure.

Some of the important factors associated with this method are as follows [4]:

- It is a very good tool for identifying undesirable events that need further investigation using the fault tree analysis method.
- The method always leaves some room to miss important initiating events.
- Usually, the method is used for performing analysis of more complex systems than the ones handled by the failure modes and effect analysis method.
- It is quite difficult for incorporating delayed recovery or success events.

Additional information on ETA is available in Refs. [4,20].

### 14.7.2 Software Sneak Circuit Analysis

This method is used for identifying software logic that causes undesired outputs. More clearly, program source code is converted to topological network trees and the code is modelled by using six basic patterns: single line, iteration loop, entry dome, trap, return dome, and parallel line. All software-related modes are modelled utilizing the basic patterns linked in a network tree flowing right from top to bottom.

The involved analyst asks questions on the use and interrelations of the instructions considered the structure's elements. The effective answers to questions asked are quite useful in providing clues that highlight sneak conditions (an unwanted event not caused by component failure) that may result in undesirable outputs. The involved analyst searches for the following four basic software sneaks:

- Wrong timing.
- Presence of an undesired output.
- The undesirable inhibit of an output.
- A program message that quite poorly describes the actual condition.

All the clue-generating questions are taken from the topograph representing the code segment and at the discovery of sneaks, the analysts conduct investigative analyses for verifying that the code does indeed generate the sneaks. Subsequently, all the possible impacts of the sneaks are assessed with care and necessary corrective measures recommended.

### 14.7.3   SOFTWARE FAULT TREE ANALYSIS (SFTA)

This method is an offshoot of the fault tree analysis (FTA) method developed in the early 1960s at the Bell Telephone Laboratories for analyzing the Minuteman Launch Control System from the safety aspect [21]. SFTA is used for analyzing software design safety and its main objective is to demonstrate that the logic contained in the software design will not cause system safety-related failures, in addition to determining environmental conditions that may lead to the software causing a safety failure [22].

SFTA proceeds in a similar manner to hardware fault tree analysis described in Chapter 4 and it also highlights software–hardware interfaces. Although fault trees for both software and hardware developed quite separately, they are linked together at their interfaces for allowing total system analysis. This is very important because it is impossible to develop software safety-related procedures in isolation, but must be considered as a part of the total system safety.

Finally, it is to be noted that although SFTA is an excellent hazard analysis method, it is quite an expensive method to use. Additional information on FTA is available in Chapter 4 and in Refs. [21,23].

### 14.7.4   PROOF OF CORRECTNESS

This is a quite useful method for performing software hazard analysis. The method decomposes a program under consideration into a number of logical segments and for each segment input/output assertions are defined. Subsequently, the involved software professional conducts verification from the perspective that each and every input assertion as well as its associated output assertion are true and that, if all of the input assertions are true, then all of the output assertions also are true.

Finally, it is to be noted that this method makes use of mathematical theorem proving concepts to verify that a given program is clearly consistent with its associated specifications. Additional information on this method is available in Refs. [17–19].

### 14.7.5   FAILURE MODES AND EFFECT ANALYSIS (FMEA)

This method was developed in the early 1950s for performing failure analysis of flight control systems and is sometimes used for performing software hazard analysis [23,24]. Basically, FMEA demands the listing of all possible failure modes of each part/component/element as well as their possible effects on the listed subsystems, system, etc.

The method is described in detail in Chapter 4 and in Ref. [23].

### 14.7.6  CODE WALK-THROUGH

This is a quite useful method for improving safety and quality of software products. The method is basically team effort among professionals, such as software programmers, system safety professionals, software engineers, and program managers. Code walk-throughs are in-depth reviews of the software in process through discussion as well as inspection of the software functionality. All logic branches as well as each statement's function are discussed with care at a significant length. More clearly, this process provides a quite good check and balances system of the software developed.

The system reviews the software's functionality and compares it with the specified system requirements. This provides a verification that all specified software safety-related requirements are implemented properly, in addition to the determination of functionality accuracy. Additional information on this method is available in Ref. [18].

## 14.8  SOFTWARE STANDARDS

There are various types of standards for use in developing software. They are useful for providing reference points to indicate minimum acceptable requirements or to compare systems [25]. Some of the standards, directly or indirectly, concerned with software safety are as follows [4,5,25]:

- DEF STD 00-55-1, Requirements for Safety-Related Software in Defence Equipment, Ministry of Defence, London, 1997.
- NSS 1740.13 Interim, Software Safety Standard, National Aeronautics and Space Administration (NASA), Washington, DC, 2000.
- IEEE 1059-1993, Guide for Software Verification and Validation Plans, Institute for Electrical and Electronic Engineers (IEEE), New York, 1993.
- IEEE 1228-1994, Software Safety Plans, Institute of Electrical and Electronic Engineers (IEEE), New York, 1994.
- IEC 60880, Software for Computers in the Safety Systems of Nuclear Power Stations, International Electrotechnical Commission (IEC), Geneva, Switzerland, 1986.
- MIL-STD-882D, System Safety Program Requirements, Department of Defense, Washington, DC, 2000.
- EIA SEB6A, System Safety Engineering in Software Development, Electronic Industries Alliance (EIA), New York, 1990.
- ANSI/AAMI, Risk Management-Part I: Application of Risk Management, American National Standards Institute (ANSI), New York, 2000.

## 14.9  PROBLEMS

1. Write an essay on software safety.
2. What are the main ways in which software can cause/contribute to a hazard?
3. Discuss at least eight basic software system safety-associated tasks.

4. Describe the software safety assurance program.
5. What are the roles a software quality assurance organization plays with respect to software safety.
6. What are the useful software safety design-associated guidelines.
7. List at least eight methods that can be used to conduct software hazard analysis.
8. What are the important factors associated with the event tree analysis (ETA) method?
9. Describe the following software hazard analysis methods:
   • Proof of correctness.
   • Code walk-through.
10. List at least seven standards that are, directly or indirectly, concerned with software safety.

## REFERENCES

1. Cha, S.S., Management Aspects of Software Safety, Proceedings of the 8th Annual Conference on Computer Assurance, 1993, pp. 35–40.
2. Hopcroft, J.E., Kraft, D.B., Sizing the U.S. Industry, *IEEE Spectrum*, December 1987, pp. 58–62.
3. Thayer, R.H., Software Engineering Project Management, in *Software Engineering*, edited by M. Dorfman, R.H. Thayer, IEEE Computer Society Press, Los Alamitos, California, 1997, pp. 358–371.
4. Mendis, K.S., Software Safety and Its Relation to Software Quality Assurance, in *Handbook of Software Quality Assurance*, edited by G.G. Schulmeyer, J.I. McManus, Prentice Hall, Upper Saddle River, NJ, 1999, pp. 669–679.
5. Dhillon, B.S., *Engineering Safety: Fundamentals, Techniques, and Applications*, World Scientific Publishing, River Edge, NJ, 2003.
6. Levenson, N.G., Software Safety: Why, What, and How, *Computing Surveys*, Vol. 18, No. 2, 1986, pp. 125–163.
7. Joyce, E., Software Bugs: A Matter of Life and Liability, *Datamation*, Vol. 33, No. 10, 1987, pp. 88–92.
8. Schneider, P., Hines, M.L.A., Classification of Medical Software, Proceedings of the IEEE Symposium on Applied Computing, 1990, pp. 20–27.
9. Gowen, L.D., Yap, M.Y., Traditional Software Development's Effects on Safety, Proceedings of the 6th Annual IEEE Symposium on Computer-Based Medical Systems, 1993, pp. 58–63.
10. Dhillon, B.S., *Medical Device Reliability and Associated Areas*, CRC Press, Boca Raton, Florida, 2000.
11. Shaw, R., Safety-Critical Software and Current Standards Initiative, *Computer Methods and Programs in Biomedicine*, Vol. 44, 1994, pp. 5–22.
12. Herrmann, D.S., *Software Safety and Reliability*, IEEE Computer Society Press, Los Alamitos, California, 1999.
13. Friedman, M.A., Voas, J.M., *Software Assessment*, John Wiley and Sons, New York, 1995.
14. Levenson, N.G., *Software*, Addison-Wesley Publishing Company, Reading, MA, 1995.
15. Keene, S.J., Assuring Software Safety, Proceedings of the Annual Reliability and Maintainability Symposium, 1992, pp. 274–279.
16. Hammer, W., Price, D., *Occupational Safety Management and Engineering*, Prentice Hall, Upper Saddle River, NJ, 2001.

17. Ippolito, L.M., Wallace, D.R., *A Study on Hazard Analysis in High Integrity Software Standards and Guidelines*, Report No. NISTIR 5589, National Institute of Standards and Technology, U.S. Department of Commerce, Washington, D.C., January 1995.
18. Sheriff, Y.S., Software Safety Analysis: The Characteristics of Efficient Technical Walk-Throughs, *Microelectronics and Reliability*, Vol. 32, No. 3, 1992, pp. 407–414.
19. Hansen, M.D., Survey of Available Software-Safety Analysis Techniques, Proceedings of the Annual Reliability and Maintainability Symposium, 1989, pp. 46–49.
20. Cox, S.J., Tait, N.R.S., *Reliability, Safety, and Risk Management*, Butterworth-Heinemann Ltd., London, 1991.
21. Dhillon, B.S., Singh, C., *Engineering Reliability: New Techniques and Applications*, John Wiley and Sons, New York, 1981.
22. Leveson, N.G., Harvey, P.R., Analyzing Software Safety, *IEEE Transactions on Software Engineering*, Vol. 9, No. 5, 1983, pp. 569–579.
23. Dhillon, B.S., *Design Reliability: Fundamentals and Applications*, CRC Press, Boca Raton, Florida, 1999.
24. Software Safety Plans, IEEE 1228-1994, Institute of Electrical and Electronic Engineers (IEEE), New York, NY, 1994.
25. Wallace, D.R., Kohn, D.R., Ippolito, L.M., An Analysis of Selected Software Safety Standards, Proceedings of the Seventh Annual Conference on Computer Assurance, 1992, pp. 123–136.

# 15 Safety in Engineering Maintenance

## 15.1 INTRODUCTION

Each year a vast sum of money is being spent worldwide for keeping engineering systems functioning effectively. The problem of safety in engineering maintenance has become a very important issue because of the occurrence of various maintenance-associated accidents throughout the industrial sector. For example, in 1994, in the U.S. mining sector around 14% of all accidents, directly or indirectly, were associated with maintenance activity [1,2]. Since 1990, such accidents' occurrence has been following an increasing trend [1].

The problem of safety in the engineering maintenance area involves ensuring not only the safety of maintenance personnel but also the actions taken by all these individuals. Engineering maintenance activities present many unique occupation-associated hazards, including conducting tasks at elevated heights or with system/equipment that has quite significant potential for releasing electrical or mechanical energy.

All in all, engineering maintenance must strive for controlling or eradicating potential hazards to ensure appropriate protection to individuals and materials, including items such as high noise levels, moving mechanical assemblies, electrical shocks, fire radiation sources, and toxic gas sources [3,4].

This chapter presents various important aspects of safety in engineering maintenance.

## 15.2 FACTS, FIGURES, AND EXAMPLES

Some of the facts, figures, and examples that are, directly or indirectly, concerned with maintenance safety are as follows:

- In 1994, around 14% of accidents in the United States mining sector were related with maintenance activity [1,4].
- In 1998, approximately 3.8 million workers suffered from disabling injuries on the job in the United States [1,5].
- In 1993, there were approximately 10,000 work-associated deaths in the United States [1,5].
- In 1979, 272 persons were killed in a DC-10 aircraft accident in Chicago because of incorrect procedures followed by maintenance workers [6].
- In 1985, 520 persons were killed in a Japan Airlines Boeing 747 jet accident because of a wrong repair [7,8].

DOI: 10.1201/9781003212928-15

- In 1998, the total cost of work-associated injuries in the United States was estimated to be approximately US$125 billion [1,4,5].
- In 1990, 10 persons were killed on the USS Iwo Jima (LPH2) naval ship because of a steam leak in the fire room, after maintenance workers repaired a valve and replaced bonnet fasteners with mismatched and wrong material [9].
- A study of safety issues concerning onboard deaths in jet fleets worldwide for the period 1982–1991 reported that maintenance and inspection was the second most important issue with 1,481 onboard deaths [10,11].
- In 1991, four workers were killed in an explosion at an oil refinery in Louisiana as three gasoline-synthesizing units being brought back to their operating state, after going through some maintenance-associated activities [12].

## 15.3   MAINTENANCE SAFETY-RELATED PROBLEMS' CAUSES AND FACTORS RESPONSIBLE FOR DUBIOUS SAFETY REPUTATION IN MAINTENANCE ACTIVITY

Over the years various causes for maintenance safety-related problems have been highlighted. Some of the important ones of these causes are shown in Figure 15.1 [4,5].

There are many factors directly or indirectly responsible for giving the maintenance activity a dubious safety reputation. Nine of these factors are as follows [13]:

- **Factor I**: Sudden need for maintenance work, thus allowing a very short time for appropriate preparation.
- **Factor II**: Frequent occurrence of many maintenance tasks (e.g., equipment failures), thus lesser opportunity for discerning safety-associated problems as well as for initiating appropriate remedial actions.

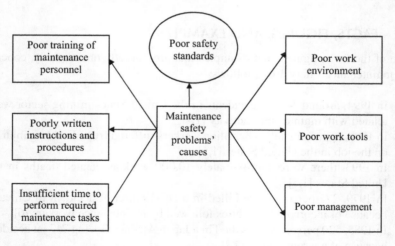

**FIGURE 15.1**   Maintenance safety problems' causes.

- **Factor III**: Performance of maintenance-related activities underneath or inside items such as pressure vessels, large rotating machines, and air ducts.
- **Factor IV**: Maintenance work conducted in unfamiliar surroundings or territory implies that hazards such as missing gratings, rusted hand rails, and damaged light fittings may go totally unnoticed.
- **Factor V**: Difficulty in maintaining effective communication with all personnel involved in the performance of maintenance tasks.
- **Factor VI**: Maintenance activities' performance in remote areas, at odd hours, and in small numbers.
- **Factor VII**: From time to time, maintenance-related activities may require conducting tasks such as disassembling corroded parts or manhandling difficult heavy units in rather poorly lit areas and confined spaced.
- **Factor VIII**: Disassembling previously operating equipment, thus conducting tasks subject to the risk of releasing stored energy.
- **Factor IX**: Need to carry heavy and rather bulky objects from a warehouse/store to the maintenance location, sometimes employing lifting and transport equipment that is way beyond the boundaries of a strict maintenance regime.

## 15.4   FACTORS THAT INFLUENCE SAFETY BEHAVIOUR AND SAFETY CULTURE IN MAINTENANCE PERSONNEL

There are a large number of factors that directly or indirectly influence safety behaviour and safety culture in maintenance personnel. For example, 24 of the factors that influence safety behaviour and safety culture in railway maintenance workers are as follows [14]:

- **Factor I**: Communication on the job (poor quality and excessive).
- **Factor II**: Feedback messages from management personnel.
- **Factor III**: Fatigue, concentration, and ability to function.
- **Factor IV**: Poor and underutilized real-time risk assessment skills.
- **Factor V**: Training methods and training needs analysis.
- **Factor VI**: Management personnel's communication methods.
- **Factor VII**: Perceived objective of the rule book.
- **Factor VIII**: Individual perception of what "safe" is.
- **Factor IX**: Supervisory personnel's visibility and accessibility.
- **Factor X**: Rule book usability and availability.
- **Factor XI**: Equipment (condition, availability, and appropriateness).
- **Factor XII**: Perceived purpose of paperwork.
- **Factor XIII**: Practical alternatives to rules.
- **Factor XIV**: Social pressure of home life.
- **Factor XV**: Competence capability and certification.
- **Factor XVI**: Safety role model behaviour
- **Factor XVII**: Volume of paperwork.
- **Factor XVIII**: Pre-job information dissemination.

- **Factor XIX**: Reporting methods.
- **Factor XX**: Physical conditions.
- **Factor XXI**: Inconsistent teams.
- **Factor XXII**: Contradictory rules.
- **Factor XXIII**: Rule dissemination.
- **Factor XXIV**: Peer pressure.

## 15.5  GOOD SAFETY-ASSOCIATED PRACTICES DURING MAINTENANCE WORK

It is very important to follow properly good practices prior to, during, and after maintenance operations because of the existence of various types of hazards. Failure to follow properly good practices during any maintenance phase can lead to potentially hazardous conditions. Four good safety-associated practices to be followed during maintenance work are as follows [15]:

- **Practice I: Prepare for maintenance during the design phase**. It basically means that preparation for maintenance actually starts during the design of the facility by ensuring that proper indicators are in place for allowing effective troubleshooting and diagnostic-related work. Furthermore, the equipment is designed so that normal safety-associated measures can easily be taken prior to the maintenance activity. More clearly, equipment is designed so that all appropriate safeguards are in place for allowing it to be drained, isolated, purged, and analyzed effectively.
- **Practice II: Prepare all staff members for maintenance operations**. Generally, maintenance activity involves opening equipment that contains hazardous material during its normal operations. Thus, it is very important to take all necessary precautions prior to working on such equipment for ensuring that it is completely free from residual material and is at a safe temperature and pressure. Often equipment is prepared for maintenance by personnel other than those actually conducting maintenance on the equipment.

  In this scenario, it is absolutely essential to prepare all involved staff members (i.e., who prepare the equipment for maintenance and the others who conduct maintenance) for maintenance operations.
- **Practice III: Highlight all potential hazards and plan effectively well in advance**. There is absolutely no substitute for proper job planning as effective equipment isolation prior to the maintenance activity starts with thorough preplanning. Also, good practice guidelines clearly state that all potential hazards are most effectively recognized during the planning process, rather than during the job execution in a quite stressful environment.

  In summary, ensure that the equipment under consideration is properly freed from all types of potential hazards and that all safety-related precautions can be satisfied effectively. In situations when procedures cannot be followed properly and/or safety-related precautions cannot be fully satisfied, do not proceed any further until a proper hazard evaluation can be conducted and a safe course of measures determined.

• **Practice IV: Plan now for the future**. This is concerned with analyzing the possible potential effects on the maintenance-related activity when changes are made to the existing process. Along with the determination of how all operations will be affected, process management personnel must carefully evaluate questions such as: Will there be need for more frequent or less frequent maintenance? Will maintenance personnel be at greater risk because of this change? and How will this change affect all the future maintenance-associated activities?

## 15.6  MAINTENANCE-ASSOCIATED SAFETY MEASURES CONCERNING MACHINERY AND GUIDELINES FOR EQUIPMENT DESIGNERS TO IMPROVE SAFETY IN MAINTENANCE

Over the years safety specialists have done much to point out various safety-related measures to be observed in working around machinery, particularly in regard to the maintenance activity. Past experiences over the years indicate that all of these as well as the application of careful planning have considerably reduced the accidents' occurrence and damage to machinery. The following maintenance-associated safety measures have proven to be quite useful [16]:

• Items such as portable electric drills, grinders, and electric motors should have proper ground wire attached for preventing maintenance workers and others coming in contact with defective wiring on machining equipment.
• All types of machines properly equipped with appropriate safety valves, alarms for indicating abnormal operating conditions, and over-speed cutouts.
• All types of electrical equipment installed as per currently approved code.
• Equipment designed for work intended should have an appropriate level of safety margin for insuring safe operation under extreme environments.
• Appropriate guards around all exposed moving parts of machining equipment.
• Ladders, platforms, and stairways with appropriate protective features.
• Safe tools for grinding and clipping and appropriate goggles for eye protection.
• Safety clothing, shoes, gloves, and hats.

Over the years, professionals working in the area of maintenance have developed various guidelines for engineering equipment designers, considered quite useful for improving safety in maintenance. Eight of these guidelines are as follows [17]:

• **Guideline I**: Install appropriate interlocks for blocking access to hazardous locations and provide effective guards against all moving parts.

- **Guideline II**: Incorporate appropriate measures/devices for early detection or prediction of all types of potential failures so that necessary maintenance can be conducted prior to actual failure with a reduced risk of hazards.
- **Guideline III**: Pay close attention to all typical human behaviours and reduce or eliminate the need for special tools.
- **Guideline IV**: Develop the design in such a way that the probability of maintenance workers being injured by electric shock, escaping high-pressure gas, and so on, is lowered to a minimum.
- **Guideline V**: Design for easy accessibility so that all parts requiring maintenance are safe and easy to check, service, replace, or remove.
- **Guideline VI**: Eliminate or reduce the need to conduct adjustments/maintenance close to hazardous operating parts.
- **Guideline VII**: Incorporate effective fail-safe designs for preventing damage or injury in the event of a failure.
- **Guideline VIII**: Develop designs/procedures in such a way that the maintenance error occurrence probability is lowered to a minimum.

## 15.7 MAINTENANCE SAFETY-ASSOCIATED QUESTIONS FOR ENGINEERING EQUIPMENT MANUFACTURERS

Engineering equipment manufacturers can play a very important role in improving maintenance safety during equipment field use by effectively addressing common problems that might be encountered during the maintenance-related activity. Questions such as the ones presented below can be very useful to equipment manufacturers in determining whether the common problems that might be encountered during the equipment maintenance-related activity have been addressed properly [17]:

- Can the disassembled piece of equipment for repair be reassembled incorrectly so that it becomes hazardous to all its potential users?
- Is the system/equipment designed in such a way that after a malfunction, it would automatically stop operating and would cause absolutely no damage?
- Was proper attention given to reducing voltages to levels at test points so that hazards to all involved maintenance workers are reduced?
- Do all the repair instructions contain effective warnings to wear appropriate gear because of pending hazards?
- Does the equipment contain proper safety interlocks that must be bypassed for conducting essential adjustments/repairs?
- Is there a proper mechanism installed for indicating when the backup units of safety-critical systems malfunction?
- Are the parts/components requiring frequent maintenance easily accessible all the time?
- Is the need for special tools for repairing safety-critical parts/components lowered to a minimum level?
- Were human factors principles properly applied for reducing maintenance problems?

- Do the instructions properly include warnings for alerting maintenance personnel of any danger?
- Does the equipment contain an appropriate built-in system for indicating that safety-critical parts need maintenance?
- Are effectively written instructions available for repair and maintenance-related activities?
- Is the repair process hazardous to all involved repair workers?
- Are all the test points located at easy to find and reach locations?
- Is there a proper system for removing hazardous fluid from the system/equipment to be repaired?
- Are warnings properly placed on parts that can shock involved maintenance workers?

## 15.8 MATHEMATICAL MODELS

Over the years, a large number of mathematical models have been developed for performing various types of reliability and availability analysis of engineering systems [18]. Some of these models can also be used for performing maintenance safety-related analysis of engineering systems. Two such models are presented below.

### 15.8.1 MODEL I

This mathematical model can be used to represent an engineering system that can either fail safely or unsafely due to maintenance error. The failed engineering system is repaired. The model is considered quite useful for predicting engineering system availability, probability of failing safely, and probability of failing unsafely due to maintenance error.

The engineering system state space diagram is shown in Figure 15.2. The numerals in boxes denote engineering system states.

The model is subjected to the following assumptions:

- All engineering system failures are statistically independent.
- Engineering system failure and repair rates are constant.
- The repaired engineering system is as good as new.

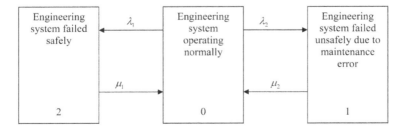

FIGURE 15.2  Engineering system state space diagram.

The following symbols are associated with the diagram:

$i$ is the ith engineering system state: $i = 0$ (engineering system operating normally), $i = 1$ (engineering system failed unsafely due to maintenance error), $i = 2$ (engineering system failed safely).

$\lambda_1$ is the engineering system constant failure rate from state 0 to state 2.

$\lambda_2$ is the engineering system constant failure rate from state 0 to state 1.

$\mu_1$ is the engineering system constant repair rate from state 2 to state 0.

$\mu_2$ is the engineering system constant repair rate from state 1 to state 0.

$P_i(t)$ is the probability that the engineering system is in state i at time $t$; for $i = 0,2$.

Using the Markov method described in Chapter 4, we write down the following three differential equations for Figure 15.2 [19,20]:

$$\frac{dP_0(t)}{dt} + (\lambda_1 + \lambda_2) P_0(t) = \mu_1 P_2(t) + \mu_2 P_1(t) \tag{15.1}$$

$$\frac{dP_1(t)}{dt} + \mu_2 P_1(t) = \lambda_2 P_0(t) \tag{15.2}$$

$$\frac{dP_2(t)}{dt} + \mu_1 P_2(t) = \lambda_1 P_0(t) \tag{15.3}$$

At time t=0, $P_0(0) = 1, P_1(0) = 0,$ and $P_2(0) = 0$.

Solving Equations (15.1)–(15.3) we obtain

$$P_0(t) = \frac{\mu_1 \mu_2}{m_1 m_2} + \left[ \frac{(m_1 + \mu_2)(m_1 + \mu_1)}{m_1(m_1 - m_2)} \right] e^{m_1 t} - \left[ \frac{(m_2 + \mu_2)(m_2 + \mu_1)}{m_2(m_1 - m_2)} \right] e^{m_2 t} \tag{15.4}$$

$$P_1(t) = \frac{\lambda_2 \mu_1}{m_1 m_2} + \left[ \frac{\lambda_2 m_1 + \lambda_2 \mu_1}{m_1(m_1 - m_2)} \right] e^{m_1 t} - \left[ \frac{(\mu_1 + m_2)\lambda_2}{m_2(m_1 - m_2)} \right] e^{m_2 t} \tag{15.5}$$

$$P_2(t) = \frac{\lambda_1 \mu_2}{m_1 m_2} + \left[ \frac{\lambda_1 m_1 + \lambda_1 \mu_2}{m_1(m_1 - m_2)} \right] e^{m_1 t} - \left[ \frac{(\mu_2 + m_2)\lambda_1}{m_2(m_1 - m_2)} \right] e^{m_2 t} \tag{15.6}$$

where

$$m_1, m_2 = \frac{-B \pm \left[ B^2 - 4\left( \mu_2 \mu_1 + \lambda_1 \mu_2 + \lambda_2 \mu_1 \right) \right]^{1/2}}{2} \tag{15.7}$$

$$B = \left( \mu_1 + \mu_2 + \lambda_1 + \lambda_2 \right) \tag{15.8}$$

$$m_1 + m_2 = -B \tag{15.9}$$

$$m_1 m_2 = \mu_2 \mu_1 + \lambda_2 \mu_1 + \lambda_1 \mu_2 \tag{15.10}$$

The probability of the engineering system failed unsafely due to maintenance error, the probability of the engineering system failed safely, and the engineering system availability is given by Equations (15.5), (15.6), and (15.4), respectively.

As time $t$ becomes very large, Equation (15.4) reduces to

$$AV_{ess} = \frac{\mu_1 \mu_2}{\mu_2 \mu_1 + \lambda_2 \mu_1 + \lambda_1 \mu_2} \tag{15.11}$$

where $AV_{ess}$ is the engineering system steady-state availability.

## Example 15.1

Assume that for an engineering system we have the following data values:

$\lambda_1 = 0.008$ failures per hour
$\lambda_2 = 0.001$ failures per hour
$\mu_1 = 0.06$ repairs per hour
$\mu_2 = 0.01$ repairs per hour

Calculate the engineering system steady-state availability by using Equation (15.11). By substituting the specified data values into Equation (15.11), we obtain

$$AV_{ess} = \frac{(0.06)(0.01)}{(0.01)(0.06) + (0.001)(0.06) + (0.008)(0.01)}$$

$$= 0.8108$$

Thus, the engineering system steady-state availability is 0.8108. In other words, there is about 81% chance that the engineering system will be available for service.

## 15.8.2   Model II

This mathematical model represents an engineering system with three states: operating normally, operating unsafely (due to maintenance problems), and failed. The system is repaired from unsafe operating and failed states. The system state space diagram is shown in Figure 15.3. The numerals in boxes denote system states.

The following assumptions are associated with the model:

- All occurrences are independent of each other.
- All engineering system failure and repair rates are constant.
- The repaired system is as good as new.

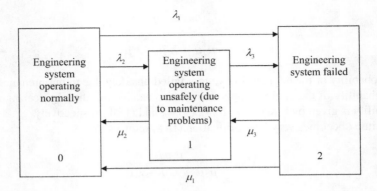

**FIGURE 15.3**   Engineering system state space diagram.

The following symbols are associated with the diagram:

  $i$ is the ith state of the engineering system: $i = 0$ (engineering system operating
    normally), $i = 1$ (engineering system operating unsafely due to maintenance
    problems), $i = 2$ (engineering system failed).
  $t$ is time.
  $P_i(t)$ is the probability that the engineering system is in state $i$ at time $t$; $i =$
    $0,1,2$.
  $\lambda_1$ is the engineering system constant failure rate.
  $\lambda_2$ is the engineering system constant unsafe degradation rate due to mainte-
    nance problems.
  $\lambda_3$ is the engineering system constant failure rate from its unsafe operating
    state 1.
  $\mu_1$ is the engineering system constant repair rate from state 2 to state 0.
  $\mu_2$ is the engineering system constant repair rate from state 1 to state 0.
  $\mu_3$ is the engineering system constant repair rate from state 2 to state 1.

Using the Markov method described in Chapter 4, we write down the following three
differential equations for Figure 15.3 [4,18]:

$$\frac{dP_0(t)}{dt} + (\lambda_2 + \lambda_1) P_0(t) = \mu_2 P_1(t) + \mu_1 P_2 \tag{15.12}$$

$$\frac{dP_1(t)}{dt} + (\mu_2 + \mu_3) P_1(t) = \mu_3 P_2(t) + \lambda_2 P_0(t) \tag{15.13}$$

$$\frac{dP_2(t)}{dt} + (\mu_1 + \mu_3) P_2(t) = \lambda_3 P_1(t) + \lambda_1 P_0(t) \tag{15.14}$$

At time $t = 0$, $P_0(0) = 1, P_1(0) = 0, and\ P_2(0) = 0$.

For a very large $t$, by solving Equations (15.12)–(15.14), we obtain the following steady-state probability equations [18]:

$$P_0 = \frac{(\mu_1 + \mu_3)(\mu_2 + \lambda_3) - \lambda_3 \mu_3}{Y} \tag{15.15}$$

where

$$Y = (\mu_1 + \mu_3)(\mu_2 + \lambda_2 + \lambda_3) + \lambda_1(\mu_2 + \lambda_3) + \lambda_1 \mu_3 + \lambda_2 \lambda_3 - \lambda_3 \mu_3 \tag{15.16}$$

$$P_1 = \frac{\lambda_2(\mu_1 + \mu_3) + \lambda_1 \mu_3}{Y} \tag{15.17}$$

$$P_2 = \frac{\lambda_1 \lambda_3 + \lambda_1(\mu_2 + \lambda_3)}{Y} \tag{15.18}$$

where $P_0, P_1,$ and $P_2$ are the steady-state probabilities of the engineering system being in states 0, 1, and 2, respectively.

Thus, the steady-state probability of the engineering system operating unsafely due to maintenance problems is given by Equation (15.17).

By setting $\mu_1 = \mu_3 = 0$ in Equations (15.12)–(15.14) and solving the resulting equations, we obtain the following equation for the engineering system reliability:

$$R_{es}(t) = P_0(t) + P_1(t)$$
$$= (Y_1 + N_1)e^{x_1 t} + (Y_2 + N_2)e^{x_2 t} \tag{15.19}$$

where $R_{es}(t)$ is the engineering system reliability at time t.

$$x_1 = \frac{-M_1 + \sqrt{M_1^2 - 4M_2}}{2} \tag{15.20}$$

$$x_2 = \frac{-M_1 - \sqrt{M_1^2 - 4M_2}}{2} \tag{15.21}$$

$$M_1 = \mu_2 + \lambda_1 + \lambda_2 + \lambda_3 \tag{15.22}$$

$$M_2 = \lambda_1 \mu_2 + \lambda_1 \lambda_3 + \lambda_2 \lambda_3 \tag{15.23}$$

$$Y_1 = \frac{x_1 + \mu_2 + \lambda_3}{(x_1 - x_2)} \tag{15.24}$$

$$Y_2 = \frac{x_2 + \mu_2 + \lambda_3}{(x_2 - x_1)} \tag{15.25}$$

$$N_1 = \frac{\lambda_2}{\left(x_1 - x_2\right)}$$                                 (15.26)

$$N_2 = \frac{\lambda_2}{\left(x_2 - x_1\right)}$$                                 (15.27)

By integrating Equation (15.19) over the time interval $\left[0,\infty\right]$, we get the following equation for the engineering system mean time to failure with repair [4,18]:

$$MTTF_{esr} = \int_0^\infty R_{es}(t)\,dt$$

$$\left[\frac{\left(Y_1 + N_1\right)}{x_1} + \frac{\left(Y_2 + N_2\right)}{x_1}\right]$$        (15.28)

where $MTTF_{esr}$ is the engineering system mean time to failure with repair.

## Example 15.2

Assume that an engineering system can be either operating normally, operating unsafely due to maintenance problems, or failed. Its constant failure/degradation rates from normal operating state to failed state, normal working state to unsafe operating state, and unsafe operating state to failed state are 0.008 failures per hour, 0.003 failures per hour, and 0.006 failures per hour, respectively.

Similarly, the engineering system constant repair rates from the unsafe operating state to normal operating state, failed state to unsafe operating state, and failed state to normal operating state are 0.004 repairs per hour, 0.007 repairs per hour, and 0.009 repairs per hour, respectively.

Calculate the steady-state probability of the engineering system being in unsafe operating state due to maintenance problems.

By inserting the given data values into Equation (15.17), we obtain

$$P_1 = \frac{(0.003)(0.009 + 0.007) + (0.008)(0.007)}{Y}$$

$$Y = (0.009 + 0.007)(0.004 + 0.003 + 0.006) + 0.008(0.004 + 0.006)$$

$$+ (0.008)(0.007) + (0.003)(0.006) - (0.006)(0.007)$$

$$= 0.325$$

Thus, the steady-state probability of the engineering system being in unsafe operating state due to maintenance problems is 0.325.

## 15.9   PROBLEMS

1. List at least seven facts, figures, and examples directly or indirectly concerned with safety in engineering maintenance.
2. List at least seven important causes of maintenance safety problems.
3. Discuss at least six factors directly or indirectly responsible for dubious safety reputation in maintenance activity.
4. List at least 20 factors that directly or indirectly influence safety behaviour and safety culture in maintenance personnel.
5. Discuss at least four good safety-associated practices during maintenance work.
6. Discuss maintenance-associated safety measures concerning machinery.
7. Discuss at least seven guidelines for equipment designers to improve safety in maintenance.
8. Write down at least eight maintenance safety-associated questions for engineering equipment manufacturers.
9. Prove Equations (15.5), (15.17), and (15.18) by using Equations (15.12)–(15.14).
10. Prove Equation (15.11) by using Equation (15.4) and assume that for an engineering system, we have the following data values:
    - $\lambda_1 = 0.006$ failures per hour
    - $\lambda_2 = 0.004$ failures per hour
    - $\mu_1 = 0.05$ repairs per hour
    - $\mu_2 = 0.02$ repairs per hour

    Calculate the engineering system steady-state availability by using Equation (15.11).

## REFERENCES

1. National Safety Council, *Accident Facts*, National Safety Council, Illinois, 1999.
2. Dhillon, B.S., *Human Reliability, Error, and Human Factors in Engineering Maintenance*, CRC Press, Boca Raton, Florida, 2009.
3. AMCP 706–132, Maintenance Engineering Techniques, U.S. Army Material Command, Department of the Army, Washington, D.C., 1975.
4. Dhillon, B.S., *Engineering Safety: Fundamentals, Techniques, and Applications*, World Scientific Publishing, River Edge, New Jersey, 2003.
5. Dhillon, B.S., *Engineering Maintenance: A Modern Approach*, CRC Press, Boca Raton, Florida, 2002.
6. Christensen, J.M., Howard, J.M., Field Experience in Maintenance, in *Human Detection and Diagnosis of System Failures*, edited by J. Rasmussen and W.B. Rouse, Plenum Press, New York, 1981, pp. 111–133.
7. Australian Transport Safety Bureau, *ATSB Survey of Licensed Aircraft Maintenance Engineers in Australia*, Report No. ISBN 0642274738, Australian Transport Safety Bureau (ATSB), Department of Transport and Regional Services, Canberra, Australia, 2001.
8. Gero, D., *Aviation Disasters*, Patrick Stephens, Sparkford, U.K., 1993.
9. Joint Fleet Maintenance Manual, Vol. 5, *Quality Assurance, Submarine Maintenance Engineering*, United States Navy, Portsmouth, NH, 1991.

10. Russell, P.D., Management Strategies for Accident Prevention, *Air Asia*, Vol. 6, 1994, pp. 31–41.
11. *Human Factors in Airline Maintenance: A Study of Incident Reports*, Bureau of Air Safety Investigation, Department of Transport and Regional Development, Canberra, Australia, 1997.
12. Goetsch, D.L., *Occupational Safety and Health*, Prentice-Hall, Englewood, NJ, 1996.
13. Stoneham, D., *The Maintenance Management and Technology Handbook*, Elsevier Science, Oxford, U.K., 1998.
14. Farrinton-Darby, T., Pickup, L., Wilson, J.R., Safety Culture in Railway Maintenance, *Safety Science*, Vol. 43, 2005, pp. 39–40.
15. Wallace, S.J., Merritt, C.W., Know When to Say "When": A Review of Safety Incidents Involving Maintenance Issues, *Process Safety Progress*, Vol. 22, No. 4, 2003, pp. 212–219.
16. Render, W.R., *Safety in Maintenance, Southern Power and Industry*, Vol. 62, No. 12, 1944, pp. 98, 99, and 110.
17. Hammer, W., *Product Safety Management and Engineering*, Prentice-Hall, Englewood Cliffs, NJ, 1980.
18. Dhillon, B.S., *Design Reliability: Fundamentals and Applications*, CRC Press, Boca Raton, Florida, 1999.
19. Dhillon, B.S., *Reliability Engineering in Systems Design and Operation*, Van Nostrand Reinhold, New York, 1983.
20. Dhillon, B.S., *Mining Equipment Reliability, Maintainability, and Safety*, Springer-Verlag, London, 2008.

# Index